SCIENCE AND PUBLIC POLICY III

ANNALS OF THE NEW YORK ACADEMY OF SCIENCES
Volume 403

SCIENCE AND PUBLIC POLICY III

Edited by Frances S. Sterrett

The New York Academy of Sciences
New York, New York
1983

JUBERT DESIGN

Library of Congress Cataloging in Publication Data

(Revised for volume 3)
Main entry under title:

Papers in science and public policy.

(Annals of the New York Academy of Sciences;
v. 368,)
Vol. 3 edited by Frances S. Sterrett.
Vols. 2- have title: Science and public policy.
Bibliography: v. 3
Includes index.
1. Science and state—United States—Congresses.
2. Medical policy—United States—Congresses.
3. United States—Social policy—Congresses.
I. Rosenberg, Bianca L. II. Sterrett, Frances S.
III. Title: Science and public policy. IV. Series:
Annals of the New York Academy of Sciences; v. 368,
etc. [DNLM: 1. Public policy—United States—Congresses. 2. Science—United States—Congresses. W1
AN626YL v. 368 etc.]
Q11.N5 vol. 368, etc. [Q127.U6] 500s [303.4'83] 81-9609
ISBN 0-89766-125-7 (v. 1)
ISBN 0-89766-126-5 (pbk.: v. 1)

SP
Printed in the United States of America
ISBN 0-89766-198-2 (Cloth)
ISBN 0-89766-199-0 (Paper)

ANNALS OF THE NEW YORK ACADEMY OF SCIENCES

Volume 403
May 26, 1983

SCIENCE AND PUBLIC POLICY III*

Editor

FRANCES S. STERRETT

CONTENTS

*The papers in this volume were presented at the 1980–1981 meetings of the Section of Science and Public Policy of The New York Academy of Sciences.

PREFACE

This third volume of *Science and Public Policy* (Volume 403 of the *Annals*), like the two previous volumes, contains papers that were presented at the monthly meetings of the Section of Science and Public Policy; the papers in this volume were presented during the academic years 1980–1981. These meetings aim to present views on topics of public policy issues to both scientifically and non-scientifically informed citizens. In today's age it becomes increasingly important to try to understand some of the raw material that influences the scientific judgement of policy makers.

Public policy issues encompass a wide range of subject matter; this volume contains substantially different topics from the previous ones. I have grouped together those papers that have a common theme.

Since Gerald Stillman's presentation stands alone, his article is printed first. He presents problems connected with energy production from municipal solid waste (garbage), and the social, economic and environmental factors associated with this technology.

In the next two papers, Evelyn Mauss and Rosalyn Yalow address controversial questions of the health effects of exposure to low-level radiation. The approach of each of these authors is quite different. For this reason, the discussions that followed Dr. Yalow's presentation, were recorded and both Dr. Mauss and Dr. Yalow were given the opportunity to edit their comments.

The last five papers deal with different problems in reference to a variety of regulations and public policies. Anita Curran discusses the problems connected with the responsibility of public health administrators to establish, interpret, and implement regulations based on controversies such as scientific facts versus public perception, economic concerns versus health concerns, public health versus personal health. John Ullmann presents current government policies on business activities to roll back recent gains in environmental protection, in safety of products and of the work place; he analyzes the related major professional, ethical and economic problems. Peter Raven-Hausen deals with scientists who, in accepting public funds to further their research, may relinquish their right to ownership and control of research data. The interpretation of their findings may consequently result in public criticism or ridicule. Ithiel de Sola Pool presents some history of regulations concerning the use of human subjects for research; he emphasizes the difference between biomedical and social-behavioral human subject regulations. The last paper is by Steven Brams who evaluates different voting systems of the political arena. He advocates a system called approval voting.

I would like to express my appreciation to the Science and Public Policy Section's Advisory Committee[a] for assistance in helping to select speakers and

[a]Members of the Advisory Committee: Neil Cowan, Marcel Faber (Vice-Chair, 1980), Phyllis Grodsky, Sally Guttmacher, Eric Holtzman, James C. King, Philip Kotlar, Evelyn A. Mauss, Seymour Melman, Bianca L. Rosenberg, Herbert J. Rosenberg, Kurt Salzinger, Steven L. Schwarcz, Philip Siekevitz, Ethel Tobach, John E. Ullmann, and Marjorie Zucker.

topics for our monthly meetings. Also I am grateful to all the authors who made presentations at the monthly meetings and prepared papers for this publication. Last but not least, thanks to Bill Boland and Joyce Hitchcock for their valuable professional and editorial advice.

Frances S. Sterrett

Municipal Solid Waste (Garbage): Problems and Benefits[a]

GERALD I. STILLMAN

Power Authority of the State of New York
New York, New York 10019

CHARACTERIZATION OF GARBAGE

A literary critic, in reviewing Dostoevski's *House of the Dead,* a novel about prison life, stated that one can judge the level of a civilization by the way it runs its prisons. Another criterion by which to judge the level of civilization is the quality of its garbage and the amount generated per person. TABLE 1 compares the residential and commercial garbage generated in three relatively affluent Long Island towns with that of New York City and the national average for the United States. The range of garbage generated per person runs from about $3\frac{1}{2}$ lb/day to 7 lb/day in the United States.

The combustible portion of garbage consists primarily of paper products, plastics, textiles and wood. TABLE 2 lists the components on an air-dried basis and a dry basis. The "air-dried basis" contains, in the example of TABLE 2, 15.4% moisture in those components that can absorb and adsorb moisture, while the "dry basis" has all moisture driven off. Adding together the first four components, one can see that paper products constitute close to 40% of the dry-weight garbage, plastics over 5%, and wood and textiles close to 8%. Nearly 30% of the dry weight is ash. The latter figure makes it clear that after incineration, nearly one-third of the original dry weight of the garbage will have to be landfilled. It should be noted, however, that the *volume* reduction is much larger, usually of the order of 10–20 times, so that only 5–10% of the original volume of the garbage would have to be landfilled.

A chemical analysis of garbage shows that, as-received, approximately one-quarter of its weight is water, another quarter is carbon, one-fifth is oxygen, and another one-quarter is inert material (which eventually becomes ash). The remainder of the as-received garbage consists of sulfur, chlorine, hydrogen, and oxygen. The amount of chlorine is generally interlinked with the amount of polyvinylchloride (PVC) contained in the plastic portion of the garbage. New plastics are being developed which may conceivably replace PVC in the next decade and thereby lessen the amount of chlorine to be found in the garbage. On an as-received basis, the particular garbage analyzed in TABLE 3 has a heat content with a higher heating value of 4370 Btu/lb. The other two columns in TABLE 3 show how the heat content per pound of garbage increases as moisture is extracted.

How does garbage compare with coal as a fuel? TABLE 4 compares

[a]This paper was presented at the October 14, 1981 meeting of the Section of Science and Public Policy of the New York Academy of Sciences.

0077–8923/83/0403–0001 $01.75/2 © 1983, NYAS

TABLE 1. Samples of Refuse Generation for Southern New York State (1980)

Location	Population	Tons/Year	(Residential & Commercial) LBS/Capita/Day
Babylon, L.I.	230,000	235,000	6
Huntington, L.I.	225,000	275,000	7
Islip, L.I.	339,000	300,000	5.25
New York City	7,400,000	6,500,000	4.75
U.S. National Avg.	—	—	3.2–4.9

TABLE 2. Composite New York City Raw Refuse Component Breakdown

Component[a]	Percent by Weight (Dry Basis)	Percent by Weight (Air-Dry Basis)
Corrugated boxes	11.69	12.73
Newspaper	5.88	7.29
Magazines, books	2.05	2.26
Other paper	19.00	22.37
Plastic, rubber	3.31	3.43
Film plastic	2.23	2.54
Textiles	4.05	4.16
Wood	3.87	4.27
Grass, green	0.67	1.19
Food waste	3.99	10.74
Smalls	3.48	4.35
Glass, stones	15.97	16.07
Metals	8.34	8.60
Total	84.53	100.00

[a]Data from air-dried refuse sampled at the New York City Greenpoint incinerator was used. An overall composite ratio of .653 to .347 for household to commercial refuse was calculated from data.

TABLE 3. Composite New York City Raw Refuse Analysis

Component[a]	Percent by Weight (Dry Basis)	Percent by Weight (Air-Dry Basis)	Percent by Weight (As Received)
Moisture	—	15.40	26.00
Sulfur	0.27	0.23	0.20
Chlorine	0.89	0.75	0.66
Carbon	33.70	28.52	24.94
Hydrogen	4.60	3.90	3.40
Oxygen	26.22	22.18	19.42
Nitrogen	0.56	0.47	0.41
Inerts	33.76	28.55	24.97
Total	100.00	100.00	100.00
HHV (Btu/lb)	5,905	4,996	4,370

[a]Values for the elemental analysis were extrapolated using the dry component breakdown from FIGURE 2 and the elemental analysis for each component provided in a New York City EPA study.

TABLE 4. Comparison of Refuse and Coal Analyses (% by weight, as received)

Component	Garbage	Eastern Coal[a]	Western Coal[b]
Moisture	26.00	2.5	28.9
Sulfur	0.20	2.3	0.5
Chlorine	0.66	—	0.0+
Carbon	24.94	75.0	48.6
Hydrogen	3.40	5.0	3.5
Oxygen	19.42	6.7	12.0
Nitrogen	0.41	1.5	0.7
Inerts (Ash)	24.97	7.0	5.8
Total	100	100	100
HHV (Btu/lb)	4,370	13,000	8,520

[a]Pittsburgh Steam Coal from *Steam*, Babcock & Wilcox, 1972.
[b]Wyoming Coal, *Electric Light & Power*, Jan. 19, 1976.

garbage with a good grade of eastern bituminous coal and a western sub-bituminous coal. Perhaps the most dramatic difference between garbage and both coals is the low carbon content: one-half as much as the western coal, and one-third as much as the eastern coal. This difference is directly reflected in the heat content of the fuels; the garbage has approximately one-half the heat content of the western coal and one-third the heat content of the eastern coal.

Another noteworthy difference is the low sulfur content of the garbage compared, particularly, to the eastern coal. With a sulfur content of 0.2% in the garbage, scrubbers to remove sulfur are not required. Yet another

TABLE 5. Composition of Municipal Solid Waste in Switzerland, New York City, and Britain

	Composition by Weight Percent % (Location and Source)					
	Switzerland			NYC	Britain	
Constituents Source:	1	2	3	4	5	6
Food waste	20	12	14.5	10.7	26	13
Textiles	4	2.5	3.0	4.2	2	2.5
Paper	36	30	33.5	41.6	37	51.5
Plastics	4	7	2	6.0	1.5	1.0
Leather and rubber	2	—	—	6.0	—	—
Wood	4	6	2.5	4.3	—	—
Ferrous and nonferrous metals	6	7	5	8.6	8.5	6.5
Street sweepings and garden waste	6	—	—	1.2	2	3
					8	
Glass	8	5	8.5	{20.4	15	
Stones, dust, and other debris	10	33.5	31		6.5	
					16	

Sources: 1. National averages as published by EAWAG (1971) (used for planning Hagenholz). 2. Municipal solid waste of Geneva (1972). 3. Municipal solid waste of Zurich (1963/1964). 4. NYC 1970, air-dry basis. 5. London (1972). 6. Birmingham (1972).

noteworthy difference is the low nitrogen content. Because this "fuel nitrogen" is low, and because the combustion temperatures in a furnace that burns garbage only are low—leading to low thermal NO_x production—garbage burning generally leads to low overall NO_x output.

TABLE 5 compares the composition of the garbage of New York City with Swiss and British garbage. The various columns in the TABLE show that, while there are some differences, by and large the composition is quite similar. The only exception is in columns 2 and 3, which show an ash component (sum of last three rows) of nearly 40% for Geneva (1972) and Zurich (1964) garbage, while the other four columns show approximately 25% ash. It is interesting to note that the paper component generally ranges between 30–40%, with the exception of column 6, which is unusually high at 51.5%.

In the last two decades, living conditions in Western Europe have improved significantly and with them the heat content of the refuse that people generate. FIGURE 1 shows the increased heat content for a number of European cities. In reading this curve, it should be noted that the heat content given is the lower heating value, which can be some 20% lower than the values given in TABLES 3 and 4. Some people, who are familiar with the early literature on garbage and garbage burning, believe that European garbage is of significantly lower quality than U.S. garbage. While this may have been true ten or more years ago, TABLE 5 and FIGURE 1 should help to dispel this view.

MAKING ENERGY FROM RAW REFUSE

There are two ways of dealing with garbage as a fuel for a combustion process. One is to process the garbage to produce a fuel that is superior to raw refuse. While there are several means of accomplishing this goal, they all involve removing moisture, ceramics, stones, glass, metals, and leaving, essentially, the combustible portion of the garbage. This results in a final product, a refuse-derived fuel (RDF), which has a significantly higher heat content than raw refuse.

The second way to use garbage as a fuel is to burn it directly, in the raw state, as it comes off the delivery truck—the mass-burning approach.

Refuse-Derived Fuel

Processed fuels can be categorized, primarily (TABLE 6) as (a) fluff RDF, (b) storable RDF, (c) pyrolysis gas, and (d) other types. The resulting RDF can be burned together with other fuels such as coal, oil, gas, or wood (depending on the form of RDF), or it can be burned in a dedicated boiler by itself.

The important characteristics of the three primary categories of RDF are shown in TABLES 7–9. Fluff and storable RDF are closely related in that storable RDF consists of fluff RDF which has been embrittled by a chemical

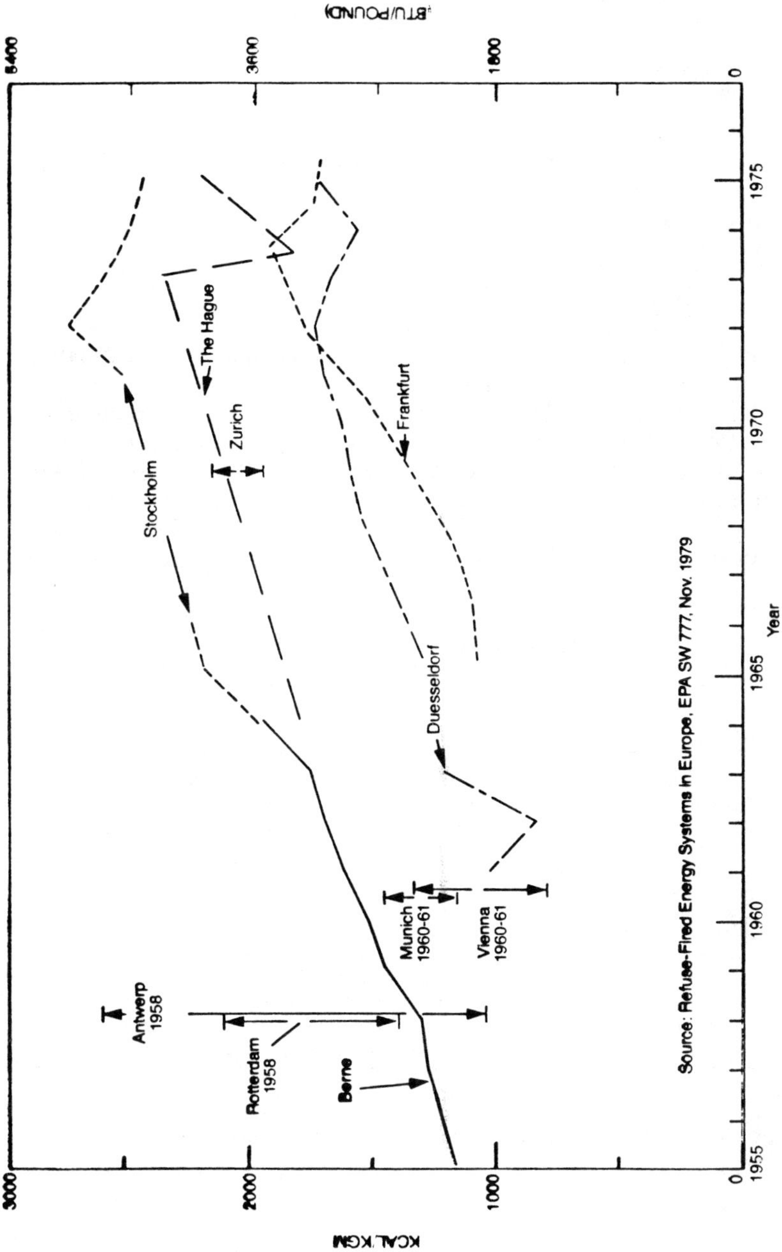

FIGURE 1. Annual average lower heating values for Berne, Stockholm, Frankfurt, the Hague, and Düsseldorf and range of values for other cities.

TABLE 6. Making Energy from Raw Refuse

Refuse-Derived Fuel

Garbage is processed to produce a "superior" fuel and to recycle salvageable materials

(a) Fluff RDF
(b) Storable RDF
(c) Gas from pyrolysis
(d) Other forms

Above processed fuels can be combusted alone in dedicated boilers or co-fired with other fuels such as coal or wood.

Direct Combustion of Raw Refuse

(a) Water-walled furnaces
(b) Refractory-lined furnaces

process and pulverized in a heated-ball mill to a consistency approximately like that of granulated sugar. Fluff RDF appears like oversized confetti. (Storable RDF is a product that was developed by Combustion Equipment Associates, called Eco-Fuel II). TABLES 7 and 8 show the increase in heat content: 6720 Btu/lb for storable compared to 4990 Btu/lb for fluff—an increase of one-third. The other quantities are also interesting. Moisture has been reduced from 28% in the fluff to 5% in the storable RDF. There is also a reduction in inerts due to the additional mechanical processing required to

TABLE 7. Fluff RDF Characteristics Summary (As-Produced Basis)

HHV	4990 BTU/LB
Moisture	28.0 percent by weight
Inerts	14.43 percent by weight
Sulfur	0.23 percent by weight
Chlorine	0.74 percent by weight
Particle size	¾–1½ in. nominal

TABLE 8. Storable RDF Characteristics Summary (As-Produced Basis)

HHV	6720 BTU/LB
Moisture	5.0 percent by weight
Inerts	11.78 percent by weight
Sulfur	0.32 percent by weight
Chlorine	0.87 percent by weight
Particle size	0.015 in. nominal

TABLE 9. Pyrolysis RDF Characteristics Summary (As-Produced Basis)

Calorific value	300 BTU/SCF
Moisture	4.02%
Inerts	125 PPM
Sulfur	300–600 PPM
Chlorine	1 PPM
Particle size	Gas

make storable RDF. The increase in the weight-percent of sulfur and chlorine in going from fluff RDF to storable RDF should not be surprising when the reduction in percent moisture by weight is taken into account; this makes the sulfur and chlorine appear to be more concentrated per pound of remaining RDF. The characteristics of the pyrolysis gas, TABLE 9, are not directly comparable to the fluff or storable RDF except for the moisture content,

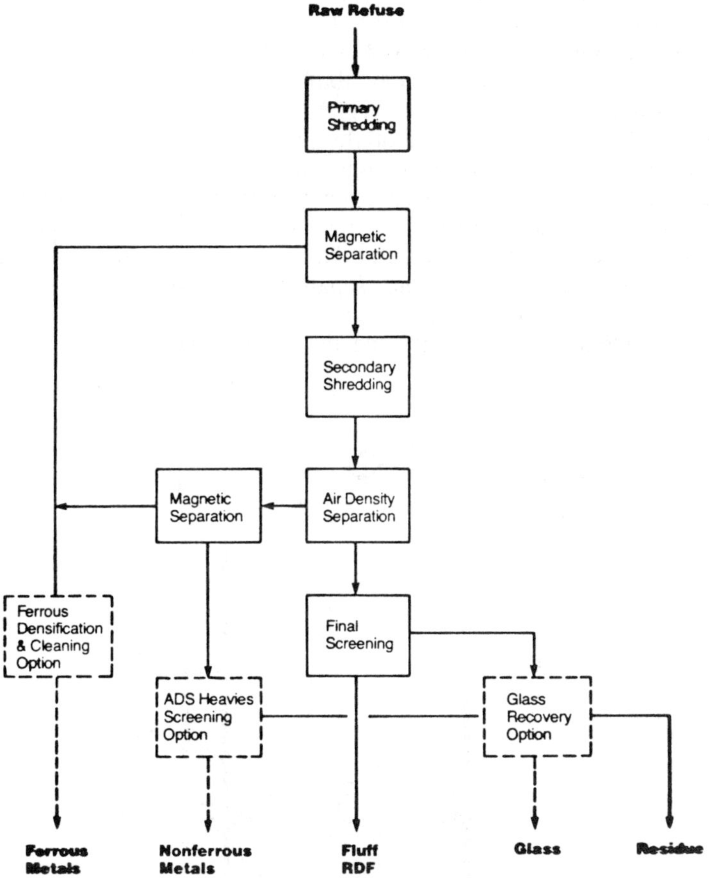

FIGURE 2. Fluff RDF—processing system schematic.

which is low—4%. The pyrolysis gas used as an example here is one produced by Union Carbide's "PUROX" process, which calls for an oxygen-blown reactor. Suffice it to say that the 300 Btu per standard cubic foot (Btu/scf) makes this medium Btu gas capable of being burned in a boiler designed for coal with practically no derating.

In the category that I labeled "other types," the most prominent one which has reached commercialization is the hydro-pulped RDF used by the Parsons Whittemore Company in the Hempstead Resource Recovery Plant (which has not been operating for nearly two years at this writing). Hydro-pulping stems from the paper industry: RDF is produced in a wet pulper (a device similar to a waste-food shredder of the type installed under a kitchen sink drain, except much larger) in a water bath so that the final product has a steady moisture content of 50% regardless of the fluctuation of moisture content in the incoming raw garbage. Still other types of RDF involve forming pellets or briquettes (for purposes of storability) of garbage that has been shredded and air classified, and sometimes mixed with other combustible material like sawdust or crushed/pulverized coal—but none has reached a commercial operating stage.

FIGURE 2 shows a schematic of the processing system to produce fluff RDF. In the particular example chosen, the raw refuse is introduced into a primary shredder. Ferrous metals are removed from the shredder output as it runs along the conveyors to the secondary shredder where the size of the material is reduced further. The output of the secondary shredding operation is conveyed to an air-classifier that separates light and heavy fractions by what is known as air density separation. The light fraction usually consists of the combustible materials—paper, plastics, light pieces of wood, textiles, for

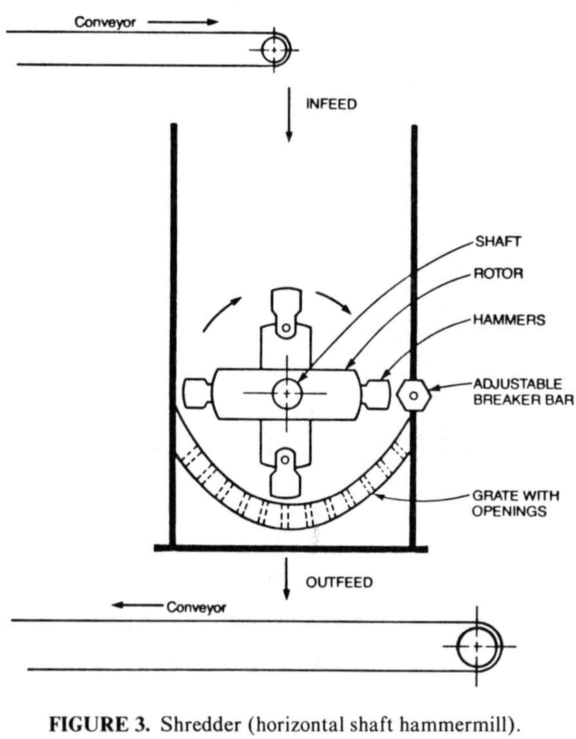

FIGURE 3. Shredder (horizontal shaft hammermill).

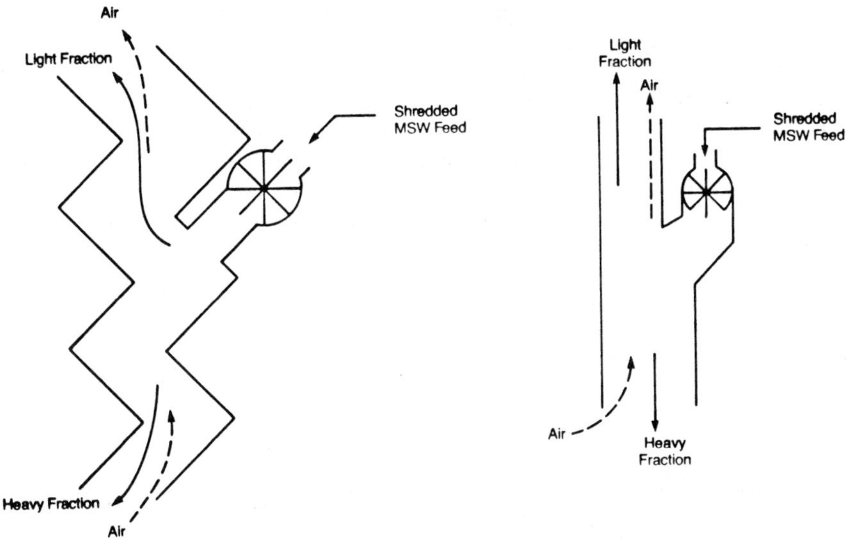

FIGURE 4. Air density separators (two vertical types).

example. The heavy fractions will contain metals, ceramics, stones, glass, bones, and other inert materials. The air classifier output is conveyed to a rotating screen that will further separate heavy from light materials. The lights become the output fluff RDF as shown; the heavy materials are either separated into glass, nonferrous metals and residues as shown, or, if such sophisticated operation is not desired, are treated as ash residue which goes to landfill.

What is a shredder? It is basically a set of rotating hammers or flails in a confined chamber that beat the garbage into smaller and smaller fragments as the shredding operation proceeds. When the fragments are small enough, they fall out of the shredder, in the example shown in FIGURE 3, through the openings of a grate or sieve, which holds back the larger fragments for additional shredding until they are small enough. The outfeed falls onto a conveyor and, if the schematic of FIGURE 2 is followed, a rotating magnet above this conveyor can pick off ferrous metals fragments.

The air-classifier usually consists of an enclosed chamber, whose shape may vary, into which air is blown in a direction counter to the flow of garbage. FIGURE 4 shows two vertical air-classifiers (or air density separators). It can be seen that the light fraction will generally flow off with the air while the heavy fraction will be able to fall against the air pressure.

A rotating screen is shown in FIGURE 5. The garbage stream is fed in at one end and travels by gravity down the sloped screen as it rotates and drops out fragments through suitably sized holes. The smaller, and lighter, fragments fall out close to the input end of the screen, while the larger, and heavier, fragments fall out toward the tail end. Clearly, the larger fragments can be put through additional processing, and, once again, fed to the screen so as to recover as much combustible material as possible.

Shredders, classifiers and screens (sometimes called trommels) are the three basic processing elements to turn raw garbage into fluff RDF. In addition, of course, there are a variety of types of mechanical and pneumatic conveyors, fans, and other material-handling equipment to transport material from one point to another within the processing factory.

All the processing activities, because they expose the moisture-bearing garbage to the ambient air within the factory, result in evaporation of some of the moisture. Each one of the processing activities is normally surmounted by a hood through which the air is sucked away to either a cyclone separation and filtration process before emission to the atmosphere, or to serve as combustion air in a furnace should there be one nearby. The air which is drawn off through

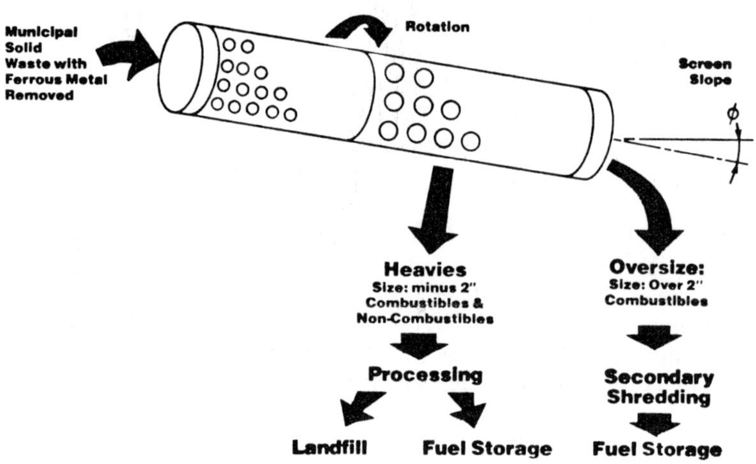

FIGURE 5. Rotary screen (trommel).

the hoods can conceivably, because of the nature of garbage, contain toxins, pathogens, and odors, which must be appropriately treated.

To get an idea of the percentage, by weight, of the material that is removed from the main RDF path by each phase of processing, see FIGURE 6 which gives the mass balance for a 3000 tpd fluff RDF processing system that operates 14 hours/day, using the other 10 hours for maintenance and repairs of the processing and materials-handling equipment. Of the 214.3 tph of raw refuse, 65% (139.3 tph) becomes the RDF which will be conveyed to the furnace for burning. A total of 14.6 tph or 6.8% of the input can be recovered as ferrous metals; 52.1 tph or 24.3% will be landfilled, and the various other fractions that can in principle be recovered are shown.

The power requirements for the processing and materials-handling apparatus are shown in FIGURE 7. The bulk of the 8086 KW required to process

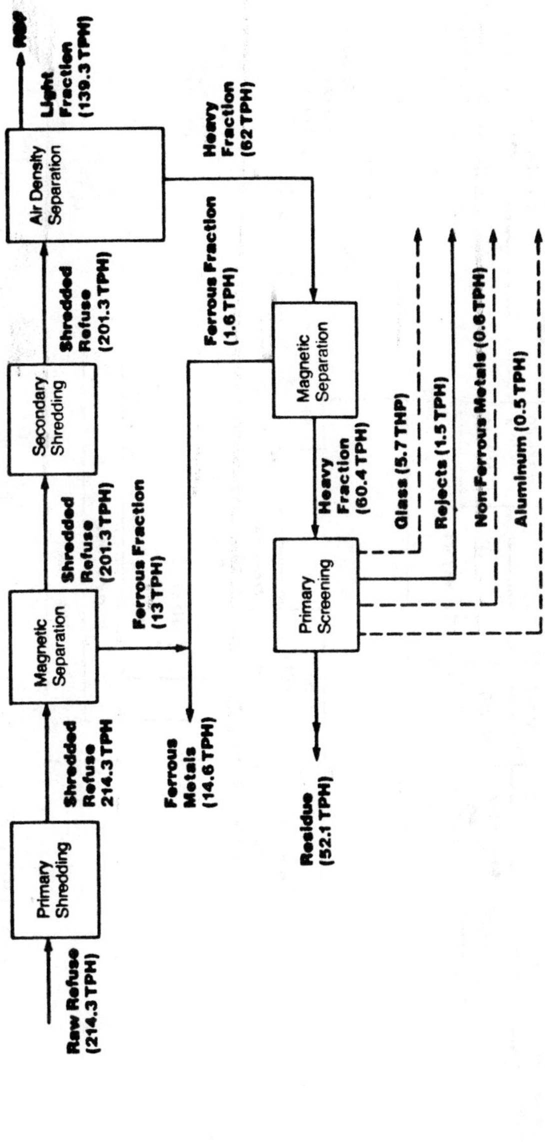

FIGURE 6. Fluff RDF processing system—Mass Balance (3000 TPD).

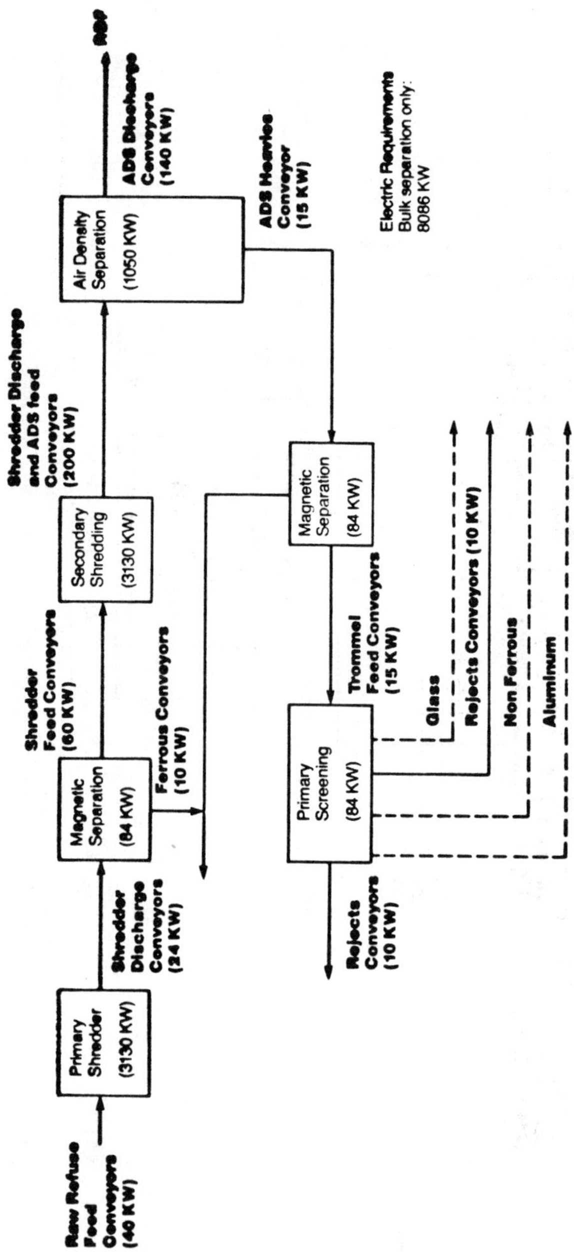

FIGURE 7. Fluff RDF processing system—power requirements (3000 TPD Plant).

214.3 tph of raw refuse into fluff RDF is consumed by three elements: the two shredders at 3130 KW each, and the air density separator at 1050 KW.

The storable RDF processing system schematic is shown in FIGURE 8 and is similar to the fluff schematic of FIGURE 2 through the secondary screening block. The output from the secondary screen is treated chemically to embrittle it and then ground it in a heated-ball mill to a consistency, as previously

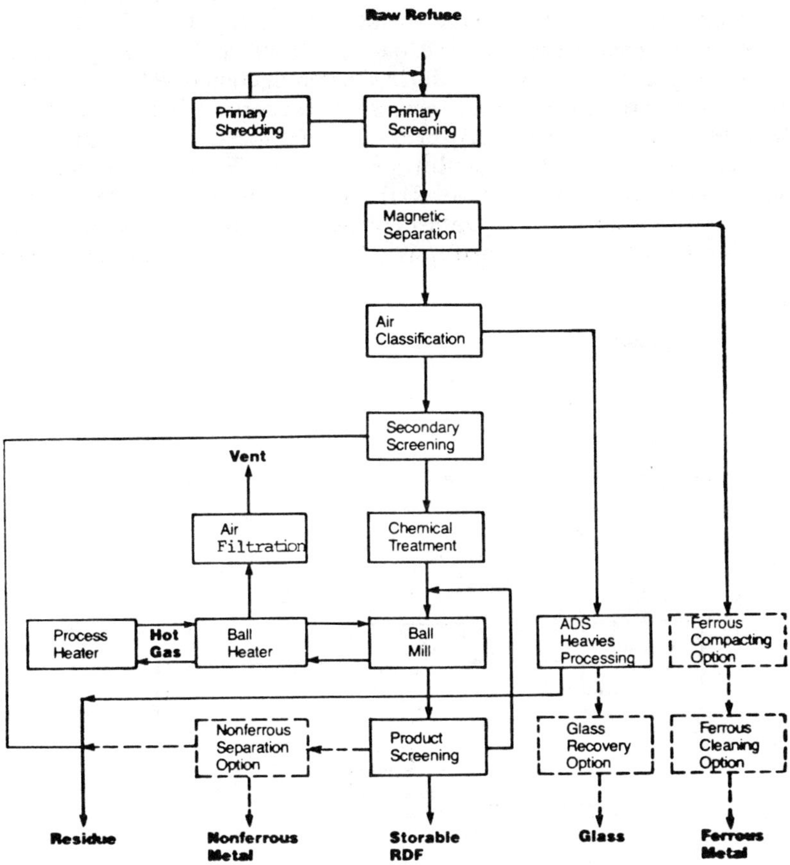

FIGURE 8. Storable RDF processing system schematic.

mentioned, like that of granulated sugar. The auxiliary processes for separation of salvageable materials are also shown.

The pyrolysis gas schematic requires two diagrams: the first, FIGURE 9, shows the shredding, air classification and screening processes which produce material that can then be transported to the pyrolysis reactor, FIGURE 10. Most of the processing that takes place in FIGURE 10 is gas cleanup so as to produce a gas that will be environmentally acceptable. It should be noted that the oxygen plant in FIGURE 10 results in a medium Btu gas of approximately 300 Btu/SCF. Were there no such plant, i.e. were air used instead of oxygen in the reactor, the final product would be a low Btu gas with a heat content of approximately 100 Btu/SCF.

Raw New York City refuse, as given in TABLE 3, is compared to fluff RDF, storable RDF and pyrolysis gas in TABLE 10. The fuel quality is compared in the left-hand portion of the matrix, the process parameters in the right-hand portion. Comparing the process parameters, the fluff RDF process has the highest energy efficiency, i.e. only about 30% of the energy contained in the incoming raw refuse is lost during the production of RDF. The residue from the RDF production plant is highest for the storable RDF (1005 tons or 33% of the incoming raw refuse) because of the more extensive mechanical processing involved in that production scheme. The smallest amount of residue comes from the pyrolysis RDF process (555 tons or 18.5% of the weight of the incoming raw refuse) which has the simplest mechanical processing (FIGURE 9). The pyrolysis RDF process is also the least energy

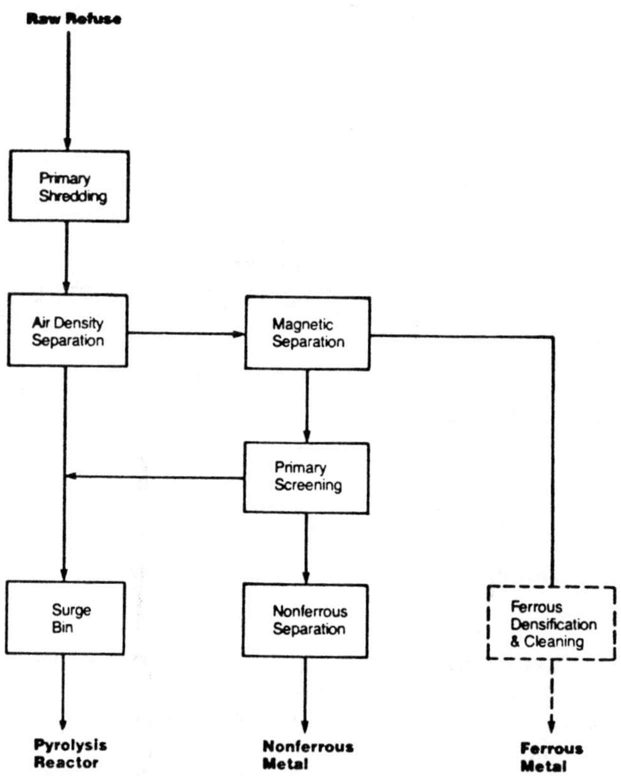

FIGURE 9. Pyrolysis gas RDF—processing system schematic (front end).

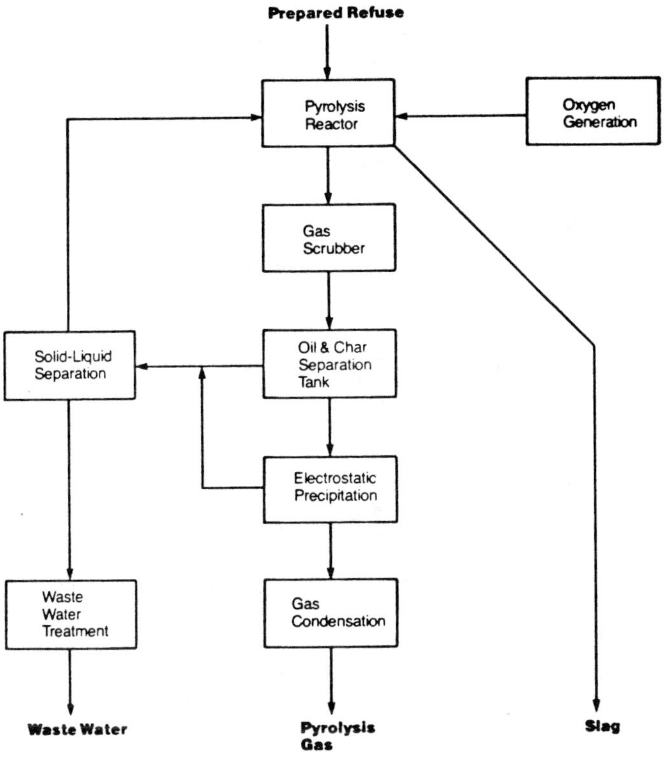

FIGURE 10. Pyrolysis gas RDF processing—system schematic (pyrolysis reactor).

efficient and, because of the need for mechanical processing, for a pyrolysis reactor plus gas and water cleanup, and for an oxygen plant, the pyrolysis scheme requires 2–2½ times the acreage as the other two schemes.

TABLE 11 provides a list of refuse processing facilities which produce several types of RDF. It can be seen that, by far, the most common process has been one to produce fluff RDF. The list is not intended to be exhaustive; it could be expanded at least twofold. All of the facilities listed have had problems of one sort or another, and many of them have shut down.

The first one on the list, in St. Louis, was a project whose objective it was to demonstrate the feasibility of cofiring fluff RDF with pulverized coal in an old Union Electric Company boiler retrofitted for that purpose. Despite a number of difficulties—both in processing the refuse into RDF and in burning the RDF in the retrofitted boiler—the demonstration was regarded as successful on the basis that the difficulties encountered all appeared to be curable. The third project on the list, in Milwaukee, inspired partly by the St. Louis experience, was an attempt at a cooperative project between the American Can Company and Wisconsin Electric Power Co. (WEPCO). American Can had developed its own fluff production system, which it called

TABLE 10. Summary Matrix for Alternative Refuse Processing Systems

		Fuel Quality					Process Parameters			
Process Technology	High Heating Value (As Received)	Sulfur Content (lb/10^6 Btu)	Chlorine Content (lb/10^6 Btu)	Inert Content (lb/10^6 Btu)	System Yields (TPD)	System Yields (10^6 Btu/Day)	Net Energy Efficiency (Percent)	Net Residue (TPD)	Magnetic Metals Recovery (TPD)	Site Requirements (Acres)
Raw NYC Refuse	4,370	0.45	1.51	57	3,000	26,220				
Ruff	4,990	0.46	1.48	28.9	1,950	19,461	69.7	820	205	12–15
Storable RDF	6,720	0.48	1.14	17.5	1,350	18,144	54.6	1,005	205	12–15
Pyrolysis RDF	300	0.14	—	0.03	52.2×10^6 (scf/day)	15,660	38.2	555	205	30–33

its "Americology" process, and it was to sell the output of its fluff production plant to WEPCO to burn together with pulverized coal in its modified Oak Creek Station. To make a long story short, it was not possible to burn the Americology fluff successfully in this station, despite numerous modifications of the Americology process.

As a matter of fact, of the entire list, only two projects can be considered as being successful: the one in Ames, IA, and the one in Madison, WI. Both of these projects cofire fluff RDF with coal in varying proportions. The Madison installation has cofired up to 11% of the heat input to the boiler with RDF. Because the heat content per pound of RDF is approximately one-half the heat content of coal, over 20% of the total fuel tonnage going into this boiler was RDF during that firing rate. The Ames facility (35 MW, tangentially fired, with an average of 10% heat input from RDF) is adding a new 65 MW boiler which will be designed to handle up to 30% heat input from RDF.

Rochester Gas & Electric has modified its Russell Station (four units ranging from 42 to 75 MW) to cofire RDF with coal at rates of up to 20% RDF heat input. The RDF will be purchased from Monroe County, which has been operating a 2000 tpd RDF factory at part load for two years, landfilling the output. This project appears to be headed for success. A much larger project is planned by the Power Authority of New York State. Construction is commencing on a 700 MW coal-fired power plant designed to accept up to 20% heat input from RDF to be produced from New York City garbage. Current plans call for this plant to be on line in 1987.

In summary, it must be said that the production of RDF and its combustion has not been convincingly successful. Implementation of the concept has been plagued with, what has in many instances turned out to be, mechanical failure of materials processing and handling equipment, and has had unforeseen problems in combusting the RDF in a boiler. This lack of success, except for the very few instances described briefly above, has given the garbage-to-energy business a bad name. It is possible that the next few years will see a greater degree of success in producing and burning RDF in boilers. In the meanwhile, attention must be turned to a technology that has been highly successful in Europe and Japan, the technology of mass-burning in which garbage is not processed but burned directly as it comes from the truck.

MASS BURNING

A simplified cross-section of a typical mass-burning energy-producing installation is shown in FIGURE 11. The garbage is unloaded in an enclosed tipping area (1) into a storage pit (2). The crane (3) mixes the garbage in the pit and accomplishes several purposes in doing so: (i) it homogenizes the garbage to a small degree; (ii) it prevents the garbage from agglomerating as it decomposes; (iii) it permits the crane operator to discover oversized objects or other undesirables which can be removed; (iv) it enhances the drying of

TABLE 11. Operating Refuse Processing Facilities

Facility	Process Employed	Design Input Capacity	Status	Remarks
St. Louis, MO	Fluff RDF	300 TPD	—R&D facility, began operating in 1973, shutdown in 1975. Planned large-scale system-dropped in 1976.	—Experienced component wear difficulties due to glass fines in refuse —Experienced plugging in pneumatic transport lines —Minimization of refuse storage time to 2 to 4 hours appeared to reduce plugging difficulties
Ames, IA	Fluff RDF	150 TPD	—Demonstration facility, operating since 1975. Has successfully burned ≈40,000 tpy. New 65 MW cofired boiler being added to fire up to 30% refuse; will use special dump grate for RDF burnout.	—Experienced component wear difficulties due to glass fines in refuse —Experienced plugging in pneumatic transport lines —Screening stage was added in 1978 in attempt to remove glass fines, however, apparently not successful
Milwaukee, WI	Fluff RDF	1200 TPD	—Commercial facility, in startup stages since 1977. Inability to burn successfully in coal-fired Oak Creek Station resulted in shutdown in 1980.	—Experienced component wear difficulties due to glass fines in refuse —Screening stage was added in attempt to remove glass fines, however, apparently not successful
Baltimore County, MD	Fluff RDF	1200 TPD	—Commercial facility, operating since 1976. Presently, output is landfilled. Possibility of cofiring with BG&E.	—Experienced component wear difficulties due to glass fines in refuse

Chicago, Southwest	Fluff RDF	1000 TPD	—Produced RDF for cofiring in 1978–79. Awaiting funding for modifying RDF plant & boiler.	—Experienced component wear difficulties due to glass fines in refuse —Required redesign of shredder and air classifier components to improve performance
E. Bridgewater, MA	Storable RDF	450 TPD	—Demonstration facility started up in 1977. Produced ECOFUEL cofiring with oil. Company in bankruptcy.	—To date only intermittent operation
Bridgeport, CT	Storable RDF	1800 TPD	—Started up in 1979. Cofired up to 50% with oil for total of 6000 tons of ECOFUEL 2 in 1980.	—Experienced excessive SO_2 emissions from boiler. RDF plant currently shut down—company in bankruptcy
San Diego County, CA	Pyrolysis RDF	200 TPD	—Demonstration plant, presently shut down.	
Charleston, WV	Pyrolysis RDF	200 TPD	—Demonstration plant, intended for test operation.	—Operation for intervals up to 3 months in 1977–78 period
Hempstead, NY	Special Fluff (wet-pulping)	2000 TPD	—Started up in 9/78. Shut down in 12/79. Generated 75×10^6 kw-hr in 1979 & sold it to LILCO.	—Odor problems —Ash handling problems —Excessive stack pollutants —Super heater burnout
Monroe Co., NY (Rochester)	Fluff RDF	2000 TPD	—Started RDF processing for landfill in 1979. To start cofiring limited amounts in early '82 in coal-burning boiler modified by RG&E.	
Madison, WI	Fluff RDF	400 TPD	—Started up in 1979. Has run successfully since. RDF cofired at up to 11% in 2-50 MW boilers.	—Uses special dump grate for RDF burnout.

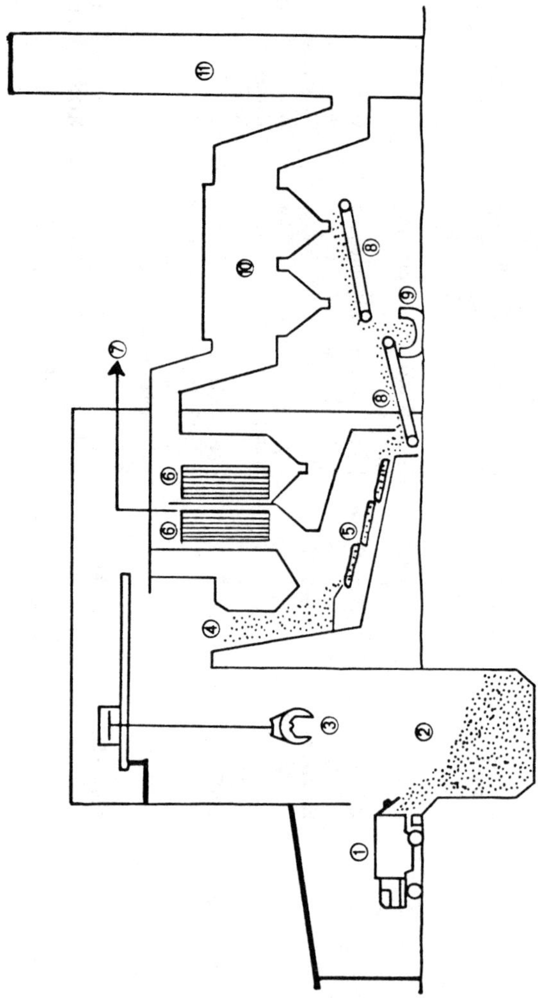

FIGURE 11. Simplified cross-section of typical mass-burning incinerator. 1. Enclosed tipping area; 2. Storage pit; 3. Crane(s); 4. Feed hopper and chute; 5. Grate; 6. Boiler convection sections; 7. Steam output; 8. Ash conveyors; 9. Ash quench; 10. Electrostatic precipitator; and 11. Stack.

garbage because of continual exposure of new surfaces to air. The crane also picks up the garbage and drops it into chute (4). The air from the enclosed tipping area (1) and storage pit (2) is continually sucked away by fans to be used as combustion air for grate (5). Thus, all this contaminated air is eventually exposed to high temperature and is deodorized and sterilized. The heated gases from above the grate pass through the convection sections of the boiler (6), heating the water in them to make steam (7) which can be used for process purposes, district heating, or generation of electricity if the pressure and temperature are high enough. The flue gases continue from the boiler section (6) through the electrostatic precipitator (10), where dust and flyash are removed, and out to the surroundings via stack (11). As the garbage burns out on grate (5), the ash falls off the right-hand end onto the ash conveyor (8) and into the quench tank (9), where it is joined by the flyash captured in the precipitators (10). The ash from the quench tank (9) is often loaded into an ash pit (not shown), whence it is periodically removed for landfilling or for other uses (such as forming part of the base material in road building).

The grate can be visualized as consisting of three major areas: the left-hand side, closest to chute (4) is the drying and ignition area; the middle section is the combustion area; and the right-hand end can be visualized as the final combustion and burnout area. The residence time of the garbage on the grate, from the time of entry from the chute until it drops off the right-hand end into the ash quench, ranges from 30 minutes to an hour typically, with a common figure being about 45 minutes.

The production of electricity places the most severe burden on a plant operator because the steam temperatures and pressures required for efficient electricity production are generally not lower than 750°F and 600 psi. The temperature and pressure requirements for process steam or district heating are much lower. The variables that the operator can control in order to produce both good burnout and the steam characteristics required for electricity production, include: grate speed; thickness of the refuse bed on the grates; the amount of underfire air; the amount of overfire air; and the ratio of underfire to overfire air. Because raw garbage is an extremely heterogeneous fuel (ranging from food and garden waste to paper to wood and plastics), it will not combust at equal rates across or along the grate, regardless of the best control efforts of the operator. Consequently, the oxidation of the carbon in the garbage (combustion is, after all, an oxidation process) will be uneven, and the flue gases leaving the grate to enter the boiler passes will consist of streamers of CO as well as CO_2, and, because of the chlorine content of the garbage, streamers of chloride gases. The CO and chlorides can attack the tubing metal and result in excessive corrosion and frequent tube replacement. The dust and flyash in the flue gases can cause excessive erosion of tube and wall surfaces. Corrosion and erosion problems have been recognized for some time; boilers of modern design can generally be operated for six months before major inspections for repairs.

Grates are of varying types. Profiles of three examples are shown in FIGURE 12. The bars of the *reciprocating* grate move from left to right and back in order to tumble and roll the garbage along as it burns out. The bars of the *reverse-acting* grate move essentially in an up and down direction, poking at the garbage to tumble and roll it down the grate. The rollers of the roller grate rotate in a clockwise direction, again, tumbling the garbage until it is burned out. In all cases, the underfire air both supplies the oxygen needed to combust the garbage and also cools the grate bars or rollers. In some plants, the underfire air is preheated by waste heat from the flue gases; while this improves the efficiency of combustion, it reduces the cooling ability of the air stream, so that good operation requires some delicate balancing.

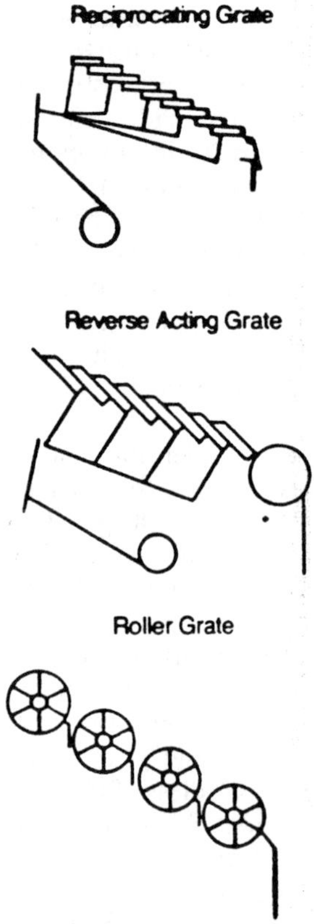

FIGURE 12. Basic types of grates for mass burning of refuse. Many variations of these basic types are available.

The furnace walls in the region over the grate (i.e. over (5) in FIGURE 11) can be constructed of water tubing which keeps the wall cool and also captures radiant heat from the flame. This type of furnace is called a *waterwall furnace* and is to be distinguished from those lined with refractory materials, which can be air-cooled, the heated air generally being used as part of the combustion air (underfire or overfire input). Until the early 1960s, almost all furnace walls were refractory lined. Today, a number of manufacturers use waterwalls only, while at least one manufacturer believes that refractory linings lead to superior maintainability and reliability.

TABLE 12. Uses for Energy

	Fire	Steam	Electricity
Sewage sludge destruction	X		
Odor destruction	X		
Large molecule destruction	X		
Electricity to grid			X
Electricity for plant auxiliary power			X
District heating		X	
Industrial process steam		X	
Destruction of exhausts from other processes	X		
Plant auxiliary power use		X	

A garbage burner produces several types of energy: the direct heat energy from the flame, steam energy, and electrical energy. Uses for each type are listed in TABLE 12.

TABLE 13 is important in that it lists the number of mass-burning plants of four manufacturers (there are other major manufacturers in addition to these, and their omission is not indicative in any way of any preference on my part) which are indicative of the essential success story of this technology. These units are located primarily in Europe, both West and East, and Japan. There are three large mass-burning plants in the United States: one in Saugus, MA (1200 tpd); one in Chicago, IL (1600 tpd, although not running at rating); and one in Harrisburg, PA (720 tpd). In addition, there are nearly 2 dozen small (200 tpd or less—mostly less than 100 tpd) mass-burning plants of domestic manufacture in the United States. These small plants produce steam only, generally for heating purposes.

The Power Authority believes that it is important to transfer the successful European and Japanese technology for larger mass-burning refuse-to-energy plants to the United States.

POWER AUTHORITY'S INVOLVEMENT IN GARBAGE-TO-ENERGY TECHNOLOGY

The Power Authority of the State of New York (PASNY) is a New York State agency which, in terms of installed capacity, is the second largest electric utility in the State, the first being Consolidated Edison. PASNY supplies about a quarter of the total load of New York City. Almost all

TABLE 13. Worldwide Inventory of Large Mass Burning Refuse-to-Energy Plants

Manufacturer	No. Plants[a] in Operation	Size Range Tons/Day	Year of 1st Plant
VON ROLL	83	150–1200	1954
MARTIN	44	120–2400	1959
VKW	35[b]	250–2800	1961
TACUMA (Japan)	27	200–1800	1971

[a]Plants consist of one or more units. A reasonable average is 2 units/plant.
[b]Plants with roller grates and energy production.

governmental electric load in New York City is supplied by PASNY; this includes state and city office buildings including the World Trade Center, the New York City subway system and commuter railroads (about 750 MW), all public schools, police stations, fire stations, city hospitals, museums, street lighting, and airports, among others. Of its 6700 MW of rated capacity, about 1800 MW is located in the New York City metropolitan area. PASNY is financed by issuing bonds (tax-exempt federally and in the state) which are paid off by the revenues from energy sales, so that no state tax monies whatsoever are used in its operations. As mentioned earlier, PASNY has begun construction of a 700 MW coal-and-refuse fired electric generating plant in New York City (on the west coast of Staten Island). Because such cofiring requires the garbage to be in the form of RDF, PASNY has an abiding interest in future advances in this technology.

PASNY, as a state agency, is working toward implementation of the Governor's 10-Point Energy Program and the State Energy Master Plan, which emphasizes the displacement of oil as a fuel (TABLE 14). Chairman

TABLE 14. Power Authority Program Objectives

Mandate
- New York State Ten Point Energy Program
- State Energy Plan
- Chairman Dyson

Objectives
- Help solve worsening land fill problem
- Displace use of oil
- Make use of indigenous "resource"

Dyson of PASNY believes also that, as a public-benefit corporation, PASNY should play a role in helping solve the worsening landfilling problem in the State. The Authority has therefore offered to finance, design, build and operate mass-burning refuse-to-energy facilities in an effort to further the transfer of successful European technology to New York State.

Such plants would be regarded as though they burned a conventional fuel (coal or oil), while recognizing that garbage is a much more difficult fuel to deal with, the point being that PASNY operators would have to develop the know-how required to successfully burn this fuel in an economical, reliable and environmentally acceptable manner (TABLE 15). In recognition of the increased maintenance and operational requirements that can be associated with the burning of a difficult fuel, it has been our philosophy to incorporate a spare boiler unit at the installations that we have proposed, so that a 700 tpd installation, for example, would consist of three 350 tpd units. This would assure reliability of incineration; if each unit has availability of 85%, i.e. it would be out of service 8 weeks per year, one would be sure of having at least two units available 95% of the time.

To implement the program, the Power Authority has made offers (TABLE 16) to New York City (South Bronx), the Town of North Hempstead, and Onondaga County/Syracuse to build and operate refuse-to-energy facilities for these communities. The response by the municipal administrations for the

TABLE 15. Program Implementation

Authority would: • finance • design • build • operate	Treat the facility as though it burned conventional fuel
Build in equipment redundancy where possible and applicable • ensure dependable and reliable operation • minimize transport of raw garbage • provide stable energy output and revenue stream	
Install most effective pollution controls	
Provide stable tipping fee • competitive with present landfill costs • escalating well below predicted inflation rates (decreased real cost of disposal)	
Prepare "project description" • for selected sites • including —site analysis —conceptual plant design —environmental considerations —socio-economic considerations —financial analysis	

first two has been positive; the third offer is being evaluated. We do not wish to get involved with more than two or three facilities at this time. In transferring technology, it is most important for us to build and operate these plants for several years so that we can be certain of the economics of doing so in the United States, and so that we can be certain also that we have mastered the knowledge to make this technology-transfer successful. Once we have convinced ourselves of the economics, we would be ready to undertake an expanded program. I can visualize this taking place around 1990.

We have been requested by towns and other administrative units in New York State to cooperate with them in solving their garbage disposal problems. While we do not wish to engage in additional plant construction and operation, we are glad to make whatever expertise we have acquired available to other administrative units in the State. In proposing our new facilities, the expertise that we have acquired above and beyond the expected understanding of the grate, boiler, electric and steam generation technology, includes site analysis, the consideration of social and economic factors, and the understanding of the importance of environmental controls of gases, liquids and solids (TABLE 17). This expertise may be of use to others.

TABLE 16. Current PASNY Involvements

SW Brooklyn Incinerator (with NYC Dept. of Sanitation)
South Bronx (with NYC Dept. of Sanitation)
Town of North Hempstead
Onondaga County/Syracuse
General Consultation Services
Arthur Kill

TABLE 17. Program Implementation

Site Analysis
- adequate size (e.g., 12–15 acres for RDF processing plant of 3000 tpd; 25 acres for 2000 tpd mass burner)
- cooling water source
- ambient air quality
- compatibility with neighborhood
- proximity to landfill or transfer station
- site tenancy (need to move anyone)

Socio-Economic Considerations
- establishing trucking pattern
- existing facilities (incinerator, shredder)
- site not on tax rolls
- proximity to existing or potential steam customers
- proximity to sewage treatment plant
- proximity to residences

Environmental Considerations
- determination of gaseous emissions particulates, SOx, NOx, CO, hydrocarbons heavy metals, pathogens, toxic agents, chlorides, fluorides
- control of emissions
- impact of emissions on ambient air quality
- liquid emissions
  —boiler blowdown, ash quench liquids, cooling tower blowdown rainwater, sanitary water
- odors and controls
- pests and vermin
- ash disposal

Mention should be made of one other significant activity in New York City. TABLE 16 lists the Southwest Brooklyn Incinerator. This is a study jointly funded with the New York City Department of Sanitation to determine the feasibility of retrofitting an existing refractory wall incinerator, south of the Verrazano-Narrows Bridge, with a waste-heat boiler to produce steam and electricity. Although at this writing the study is incomplete, indications are that such a retrofit makes economic sense. It is our hope that we and the City will find a way to cooperate and to implement the actual retrofit of this incinerator as well as other incinerators which exist in the City.

New York City generates some 22,000 tpd of garbage with a potential for conversion to energy. The City itself is currently proposing to build a 3000 tpd garbage-to-steam facility in the Brooklyn Navy Yard. The Power Authority, with its cofired plant at Arthur Kill and its proposed mass-burning plant in the South Bronx, will help to dispose of nearly 5000 tpd of garbage. If existing City incinerators, currently shutdown, are converted to energy production, following the example of the study for Southwest Brooklyn, another 2000 tpd of garbage may be accounted for. Thus, between the City and the Power Authority, some 10,000 tpd of garbage, currently being landfilled, will be incinerated to produce electric or steam energy. Add to this another 3000 tpd which will continue to be disposed of in currently operating incinerators, the City could have its garbage problem more than half licked before this decade is over.

Health Effects of Ionizing Radiation in the Low-Dose Range[a]

EVELYN A. MAUSS

Department of Physiology
College of Dentistry
New York University
New York, New York 10003

INTRODUCTION

The health effects of low-dose ionizing radiation on human populations have been difficult to establish. The question has generated heated controversy, which numerous studies have still not resolved; The National Academy of Sciences Advisory Committee on the Biological Effects of Ionizing Radiation, charged with revising its 1972 report (BEIR II),[1] issued a new report last year (1979) that was withdrawn because of profound disagreements within the Committee. The importance of pursuing the question lies in the need to enunciate limits of permissible exposure in the work place and in the general environment (from medical/dental radiation and nuclear reactor/military technologies), as well as in our collective wish to have answers whenever possible. In this presentation I shall state the problems of estimating the health effects of low-level ionizing radiation, describe or discuss selected contributions to the literature on somatic and genetic risks (referring also to a few large studies that are in the design or data collection stage), and consider briefly whether revision of present risk estimates is indicated.

THE PROBLEMS

The two major problems of assessing specifically the cancer risks of low-dose ionizing radiation—the principles are also applicable to genetic effects—have been elucidated by Land.[2,3] The first is the need for very large sample sizes if precise direct estimations of excess risks are sought. "This is because the precision of an estimate should be a few percent of the estimated value, and because the sample size necessary to attain that precision increases disproportionately as the true risk becomes smaller." Thus if the excess risk is proportional to dose, and if the population necessary to determine the effect of

[a]**Editor's note:** For its November 5, 1980 meeting, the Section of Science and Public Policy had announced that Dr. Arthur Tamplin, then of Natural Resources Defense Council, Inc., would speak on biological effects of ionizing radiation in the low-dose range. On the very night of the meeting, it became clear that Dr. Tamplin would not be able to give his talk. Dr. Evelyn A. Mauss agreed to serve as substitute speaker, drawing partly on a literature review she had prepared for another purpose and partly on an Information Bulletin Dr. Tamplin had written for NRDC. The present paper is a synthesis of the two, approximately as it was delivered at the meeting.

0077-8923/83/0403-0027 $01.75/2 © 1983, NYAS

a 100 rad[b] exposure is thought to be one thousand, the sample size necessary to determine the effects of 10 rad might be not ten thousand, but one hundred thousand, and the number necessary if the dose were one rad could be about ten million. On this basis, it is the view of some statisticians that the sample sizes necessary to obtain conclusive data are too large to be practicable and that direct observation as the way to answer the question is precluded. However, several investigations[4–6] suggest that certain new data and novel analyses of older data may provide useful direct observation answers.

The alternative to direct observation is extrapolation to low doses from high dose effects, for which there are a few well-observed populations (e.g., survivors of the atom bomb, children treated with radiation for thymic enlargement, adults treated with radiation for ankylosing spondylitis). The problem with this approach is that several kinds of curves can be fitted to the available data. The choice of curve is crucial: if a linear dose-response relationship pertains, the interpretation must be that there is no safe level of radiation exposure; that is, there is no threshold below which radiation may be presumed to be safe. This is the assumption that has informed most decision-making about permissible radiation exposures including BEIR II and the majority view in the first BEIR III. A quadratic or linear-quadratic curve would be consistent with the interpretation that effects at low doses are smaller than has been generally assumed, a "position" that has some strong proponents. But other shapes, with level or rising slope at low doses, appear to fit the data for particularly vulnerable subgroups in the population[7] and prompt the speculation that effects at low doses are significantly underestimated by linear extrapolation.

The uncertainties of present radiobiologic theory and of precise knowledge of effects of different dose rates, qualities of radiation, and irradiated sites, make it necessary, when low-dose effects are extrapolated from high-dose data, to choose between "imprecise, and therefore useless, estimates of risk from low-dose radiation using a general model, and precise, but possibly misleading, estimates using arbitrarily oversimplified models."[2] Estimates can be and have been made, but the problems of applying them to the practical purposes of public policy-making are formidable.[c]

[b]The amount of energy absorbed from radiation by a mass of material is measured in *rads* (100 ergs per gram).

The absorbed dose in rads multiplied by a factor that takes into account the differing biological effects of different types and qualities of radiation is the dose (roentgen) equivalent for man, expressed in *rems*. This unit is widely used, especially in estimating exposures to more than one type of radiation.

The unit for population exposure is the *person-rem*. It is the sum of the dosages to the individual members of the population. The exposure of 1 million individuals to 1 rem equals 1 million person-rem, and the exposure of 10 million individuals to 0.1 rem is also equal to 1 million person-rem.

[c]**Note added while manuscript was in preparation:** Land's studies doubltess figured in the one most-publicized recommendation of the Comptroller General's report to the Congress of the United States, issued in January 1981,[8] that in future Federal research on health effects of ionizing radiation exposure, increased priority and emphasis should be given to studying the mechanisms of cancer induction through cellular or molecular studies and other fundamental research, and less to epidemiologic studies. As might have been expected, this recommendation has been criticized.[9]

SOMATIC EFFECTS

In 1962, McMahon *et al.*[10] reported that children born after their mothers had had one to six pelvic radiographs (average dose per radiograph: one rad) were about 42% more likely to die of cancer in the first 10 years of life than were children not irradiated *in utero*. The directly proportional dose response relationship agreed in direction and degree with an earlier estimate by Stewart *et al.*,[11] whose paper had been criticized for the presence of possible biases. Other studies, for example, that of Graham *et al.*,[12] have supported the findings of estimated relative risk of approximately 1.5. In an analysis that contributed constructively to this issue, Mole[13] compared singleton and twin births in which radiography rates of 10 and 55%, respectively, were related not to specific medical indications but to the suspicion of twinning. The death rates from leukemia and solid cancers in the irradiated singletons and twins were similar, and were higher than in children with no history of *in utero* x-ray exposure.

In an "examination of the evidence" on prenatal (and other diagnostic) radiation and malignancy, Oppenheim and co-authors[14] concluded—on the basis of their identification of design defects and biases—that exposure to diagnostic x-rays "may be less harmful than has been claimed." The consensus, though, is that a strong association between increased risk of leukemia and prenatal exposure to diagnostic x-rays has been demonstrated.[15]

In 1979 Lyon[16] reported that excess childhood leukemia in southern Utah appeared to be associated with fallout from 26 above-ground nuclear weapons tests in Nevada between 1951 and 1958. The observed leukemia mortality was 2.44 times the expected leukemia mortality for children living in high fallout counties as compared to control children from before and after the test period. Lyon's work was analyzed and criticized by Department of Energy (DOE) and General Accounting Office (GAO)[8] statisticians. Lyon's responses to these criticisms, as well as his new "further information" paper (with co-authors),[17] indicate that children exposed to radioactive fallout have experienced a higher than "spontaneous" incidence of leukemia.[d]

Another body of data that has been a source of controversy, both scientific and political, was gathered during the '60s and '70s and presented by Mancuso and co-workers in the late '70s.[5,18] In this work, data were collected from the employment, radiation badge, and social security records of 27,960 workers at the Hanford Works in Richland, Washington. The amounts of radiation exposure ranged from 0 to 50 rads over the time of employment. These exposures were estimated to have caused 28.6 cancer deaths (about 5%). The estimate implies that the Hanford workers' risk of dying from radiation-induced cancer is perhaps ten times greater than expected under usual risk estimates. The implications of this for the nuclear industry and

[d]**Note added in proof:** The controversy generated by Lyon's work may yet be resolved: as of May 1982, the National Cancer Institute has funded a $6.6 million proposal by the University of Utah to evaluate the long-term effects of exposure to radioactive fallout in children and adults in the states of Utah, Arizona, and Idaho. Dr. Joseph Lyon is the principal investigator.

present radiation safety standards are serious, and are thought to have motivated the allegedly punitive treatment to which Mancuso was subjected.[19] The Mancuso-Stewart-Kneale data have been critically analyzed by Land[2] and Gilbert and Marks,[20] and further explicated by the authors.[21]

A study by Linos *et al.*[22] fails to show any association between low-level diagnostic radiation and leukemia. This is a case-control study of 138 patients with leukemia and 276 controls. All the cases occurred in residents of the same county. The exposure data—amounts of radiation were less than 300 rads to the bone marrow—are based on detailed medical records at the Mayo Clinic and another local medical practice. For each radiation dose for both myelocytic and lymphocytic leukemia, the estimated relative risks were usually less than one, and when this value was exceeded it did not approach statistical significance. The authors conclude "that low levels of exposure to medical radiation most probably did not increase the risk of leukemia in this community, but that if they did, the factor of increase is almost surely less than 2.0."

Many analyses of low-dose radiation effects have utilized data from a pioneer work called the Tri-State Leukemia Survey. This study was carried out in specific areas of New York, Maryland, and Minnesota between 1959 and 1962. There were about 1700 leukemia cases in the study area over the 3 years, about 300 of which were in children. The stratified random control sample that was selected provided a sampling ratio of approximately 1 person in 3000. The population base was about 13,000,000. Data collection took 4 years, data analysis another several. Actually, the latter has continued almost to the present.

In addition to prenatal radiation and cancer mentioned above,[12] and preconception radiation effects to be described below,[7,12] analysis of the Tri-State Survey data has suggested a relationship between diagnostic radiation and adult male leukemia,[23] and has called attention as well to the possibility that there are high risk groups in the population who may be especially vulnerable to radiation.[24] More recently, Bross and co-workers[4] constructed a dose response curve within the one rad range from data on men exposed to "ordinary diagnostic radiation" and concluded that evidence previously obtained by extrapolation from high-dose to low-dose levels may underestimate the health hazards of diagnostic radiation by an order of magnitude. This paper was critically reviewed[25] at the time of its appearance. Complex statistical questions are at issue and have not been resolved.

As did Mancuso and co-workers,[18] Johnson, in his studies of workers at the Rocky Flats (nuclear weapons) Plant in Jefferson County, Colorado, found shifted ratios with respect to the kinds of cancers in their study populations.[26] The Rocky Flats Plant has employed up to 10,000 workers since it started operation in 1959 and there have been indications of increased morbidity not only in the workers but in county residents in the proximity of the plant. While there have been annual reports of morbidity, mortality, and radiation exposure information on workers still active or retired and receiving benefits, the data have never been adequate for comprehensive routine epidemiologic surveillance of the employees. In 1979 the DOE started a long-term epidemiologic study of the transuranium workers in the Los Alamos Scientific Laboratories in which Rocky Flats workers are included. Meantime, Dr. Carl Johnson, Director of the Jefferson Health Department, analyzed the limited Rocky Flats data that were available as of July 1980. A comparison of the proportionate morbidity ratios (pmr) for the workers in the period 1951 to 1979 with those for white males in the state of Colorado in the period 1969–1971, age-adjusted, reveals differences between the two populations. Thus, "if the Rocky Flats workers had had the same experience with cancer as all white males in Colorado, cancer of the brain would account for 1.3% of all cancers. However, cancer of the brain actually accounted for 10.2% of all cancers, about eight times greater than expected. In this group of 3937 employees we could expect to find 1.6 cases of brain tumor, but actually found 13 cases ($p \leq 0.05$) of which 13 have now died." The observed pmr for malignant melanoma was 4.7%, nearly three times as great as expected. The pmr for cancer of the lung was slightly greater than expected, and for other cancers the pmrs were smaller or similar. The reasons for thinking these results may be related to exposure to plutonium and other radioactive gases and particulates in the plant and its environment are presented.

Military personnel who were present at one or more atomic weapons tests in Nevada during the '50s—the total number is estimated at 250,000—have been suggested for careful, large-scale study of radiation exposure effects. Test "Smoky" specifically has been of interest because of an excess of leukemia cases among 3224 men who participated in it.[27,28] The Veteran's Administration has rejected the appeal of these veterans or their survivors to recognize the cancers as service-related, but Congressional interest has resulted in review of the issue and authorization for studies by five Federal agencies.

Another population that may yield useful information are nuclear shipyard workers. Observation of a cluster of leukemia cases and rumors of others among Portsmouth Naval Shipyard workers led Najarian and Colton[29] to seek the appropriate mortality data. Analysis of 1722 deaths between 1961 and 1967 indicated that workers who had been employed maintaining nuclear vessels had higher death rates from leukemia, other hematologic cancers, and all cancers than workers employed in non-nuclear occupations. The determination of exposure to radiation on the job was made by next-of-kin of the decedents. When Navy employment records of shipyard occupations were used and the data reanalyzed,[30] the ratios of observed to expected deaths from cancers were still higher than one, but were not statistically significant. The dose-response relationship based on a proportional mortality analysis showed that for hematologic cancers there was a clear trend toward increasing ratios. The workers in six nuclear naval installations including the Portsmouth Naval Shipyard are being studied intensively now by a team headed by Matanoski at The Johns Hopkins University School of Hygiene and Public Health. The study, heavily funded by the Federal government, has been in the design or data collection stage with no new or definitive findings yet reported.

In addition to causing increased risk for various cancers, low-level ionizing radiation may be implicated in excessive risk of death from heart disease, according to a report to the 1980 International Symposium on Cancer by Matanoski.[31] Another somatic effect, reported but not verified in man, is accelerated aging.[32]

GENETIC EFFECTS

There have been few studies of genetic or preconception effects of exposure to low-level ionizing radiation in human populations. Genetic effects have been sought and could most obviously have been expected in the offspring of atomic bomb survivors, but positive findings have not yet been reported. This may be due, among other factors, to masking of mutations, delayed expression of them, incomplete data on spontaneous abortions, self-selection out of the pool by an undetermined number of survivors, or perhaps the genetically atypical nature of the survivor population (as suggested by Rotblat[33]).

Populations in areas with higher than usual background radiation have been investigated for the presence of chromosomal aberrations and developmental anomalies. The former have been found in high radiation areas in Brazil[34] and in Kerala, India.[35] The latter have been described in the study by Kochupillai *et al.*[35] In Kerala villages with high background radiation (1500–3000 mr yr^{-1}), Down's syndrome and mental retardation were markedly more prevalent than in villages with lower radiation (ca. 100 mr yr^{-1}). The two populations—12,918 and 5938, respectively—were similar with respect to an array of variables, but the "prevalence of severe mental retardation of genetic origin was four times higher in the study than in the control population, whereas that due to acquired perinatal and postnatal causes is comparable."

Gentry *et al.*,[36] in a paper on congenital malformations in New York State, found elevated rates of congenital malformation in the counties with relatively high concentrations of radioactive elements. This coincidence—higher background radiation with higher rates of congenital malformations—pertained regardless of socioeconomic status, ethnicity, consanguinity, utilization of home gardens, history of rubella, stillbirths, medical/dental radiation, etc. The demographic data and duration of residence histories suggested to the authors that the effects could have been attributable to preconception as well as prenatal radiation exposure.

Using the Tri-State Survey data base, Graham *et al.*[12] reported a 60% excess of leukemia in children whose mothers received diagnostic radiation up to 10 years before conception and a 30% excess for preconception paternal diagnostic x-ray exposures.

Bross and Natarajan[7,37] also reported genetic damage from diagnostic radiation, but concluded that for an especially susceptible subgroup in the general population of children, the risk of leukemia was many times (i.e., not just 1.5 times) increased. The subgroup of children was defined by the presence of certain "indicator diseases." In these studies, the Tri-State Survey data were utilized again, the analyses now being based on a mathematical model that described the population at risk and embodied the authors' hypothesis that radiation exposure produced both the indicator diseases and the leukemia. The relative risks were estimated to be 25 for leukemia and 5 for the indicator diseases (as contrasted to the usually calculated relative risk of approximately 0.5—i.e., increased risk of about 50%). The work of Bross *et al.* has been critically—both negatively and positively—reviewed.[38,39]

In a very recent study, Shiono *et al.*[40] found that the relative risk of childhood neoplasia following preconception x-ray exposure was higher than 2, an effect somewhat greater in their study population than that caused by exposure during pregnancy.

RISK ESTIMATES[e]

Estimates of the health risks for populations and individuals who are exposed to low-dose ionizing radiation are inexact because of statistical and other methodologic problems stated above. In addition, estimates of cancer induction depend upon whether the calculators use the absolute or relative risk model.

The relative risk model is based upon radiobiologic data that indicate that the rate of induction of various cancers per rem is proportional to the spontaneous incidence rate. At younger ages when the spontaneous cancer death rate per year is low, the rate of induction per year is also low, but at older ages when the spontaneous cancer death rate is higher, so is the rate of radiation-induced cancers. The absolute risk model tends to average these changing rates into a single value. (If the data base were complete, both models should theoretically give the same risk estimate.)

The most widely accepted risk estimates, those of the 1972 BEIR Report, are based on both risk models and on the linear hypothesis.

In an Information Bulletin prepared for the National Resources Defense Council,[41] Tamplin evaluates the BEIR Report figures, calculated by the Committee to be 100 to 450 induced cancers per million person-rem. Tamplin asserts that in deriving these estimates, inadequate consideration was given to the effects of *in utero* exposure. "The BEIR Report assumes that the risk associated with *in utero* exposure extends only throughout the first 10 years of life. If the risk from *in utero* exposure were assumed to extend throughout the lifespan, the risk estimate would be doubled and the risk of cancer would become 1000 cancers per million person-rem."

Tamplin finds also that the Mancuso-Stewart-Kneale figures for the Hanford installation workers,[5] the study by Pochin of leukemia induction in atomic bombs victims exposed to 0–9 rads at the time of the bombings,[6] the work by Rotblat on the early entrants into Hiroshima,[42] and the reports by

[e]**Note added in proof:** It is not known at this time whether the recently revised estimates of radioactive fallout from the Hiroshima and Nagasaki bombs[45] will have negligible or significant impact on risk estimates for cancer induction. There is widespread and urgent interest in having the Federal government provide prompt and adequate support for completion of the relevant research.

Lyon and associates of children in Utah who had been exposed to fallout from nuclear weapons[16,17] lend "strong support" to the conclusion that the linear hypothesis underestimates the effect of low-dose/low-dose-rate irradiation by a significant factor.

Other scientists, among them Karl Z. Morgan,[43] have expressed reservations about the alleged conservatism of the linear hypothesis.

With respect to risk estimates of genetic effects, Tamplin describes a reanalysis[44] of mice data on which BEIR Report estimates were based, and concludes that the BEIR figures of 30 to 750 genetic disorders per million person-rem may also be a significant underestimation.

CONCLUSION

Although it is now generally conceded that there is no safe level of radiation exposure, estimation of the precise effects of low-level exposure remains inconclusive. The studies on which conclusions must be based are for the most part disparate, not prospective, and not replicable. The weight of the evidence suggests that the hazards are real, and that current and conventional risk figures underestimate the effects of exposures in the low-dose range (see note *d,* p. 29).

REFERENCES

1. Advisory Committee on the Biological Effects of Ionizing Radiation. 1972. The Effects on Populations of Exposure to Low Levels of Ionizing Radiation. National Academy of Sciences/National Research Council. Washington, D.C.
2. Land, C. E. 1981. Estimating cancer risks from low doses of ionizing radiation. Science **209:** 1197–1203.
3. Land, C. E. 1979. Estimation of Risk from Low-dose Exposures to Ionizing Radiation. Interagency Task Force on the Health Effects of Ionizing Radiation: Report of the Work Group in Science, Appendix 4. U.S. Dept. of Health, Education, and Welfare, Washington, D.C., June.
4. Bross, I. D. J., M. Ball & S. Falen. 1979. A dosage response curve for the one rad range: Adult risks from diagnostic radiation. Am. J. Public Health **69:** 130–136.
5. Mancuso, T. F., A. Stewart & G. Kneale. 1977. Radiation exposures of Hanford workers dying from cancer and other causes. Health Phys. **33:** 369–385.
6. Pochin, E. E. 1976. Radiology now: Malignancies following low radiation exposure in man. Brit. J. Radiol. **49** (583): 577–579.
7. Bross, I. D. J. & N. Natarajan. 1977. Genetic damage from diagnostic radiation. J. Am. Med. Assoc. **237** (22): 2399–2401.
8. Report to the United States Congress by the Comptroller General. 1981. Problems in Assessing the Cancer Risks of Low-level Ionizing Radiation Exposure. Vol. 2. U.S. General Accounting Office, Gaithersburg, MD.
9. Johnson, C. J. Re: Research on radiation dangers. Letter *in* The Nation's Health, April 1981.

10. MacMahon, B. 1962. Prenatal X-ray exposure and childhood cancer. J. Natl. Cancer Inst. **28:** 1173–1180.
11. Stewart, A., J. Webb & D. Hewitt. 1958. A survey of childhood malignancies. Brit. Med. J. **1:** 1495–1508.
12. Graham, S., M. L. Levin, A. M. Lilienfeld, L. M. Schuman, R. Gibson, J. E. Dowd, L. Hempelmann. 1966. Preconception, intrauterine, and postnatal irradiation as related to leukemia. Natl. Cancer Inst. Monogr. **19:** 347–371.
13. Mole, R. H. 1975. Ionizing radiation as a carcinogen: Practical questions and academic pursuits. Brit. J. Radiol. **48** (567): 157–169.
14. Oppenheim, B. E., M. L. Griem & P. Meier. 1975. The effects of diagnostic X-ray exposure on the human fetus: An examination of the evidence. Radiology **114:** 529–534.
15. Interagency Task Force on the Health Effects of Ionizing Radiation. 1979. Report of the Work Group on Science. U.S. Dept. of Health, Education, and Welfare, Washington, D.C.
16. Lyon, J. L., M. R. Klauber, J. W. Gardner & K. S. Udall. 1979. Childhood leukemias associated with fallout from nuclear testing. N. Engl. J. Med. **300:** 397–402.
17. Lyon, J., J. Gardner, D. West & L. Schussman. 1980. Further information on the association of childhood leukemias with atomic fallout. *In* Banbury Report No. 4, Cancer Incidence in Defined Populations. J. Cairns, J. L. Lyon & M. Skolnick, Eds. Cold Spring Harbor, NY. pp. 145–162.
18. Kneale, G., A. Stewart & T. Mancuso. Reanalysis of data relating to the Hanford study of the cancer risks of radiation workers. Presented at the International Atomic Energy Agency Meeting, Vienna, Austria, March 13–17, 1978.
19. Hines, W. 1979. Press coverage of low-level radiation. *In* Public Control of Environmental Health Hazards. Ann. NY Acad. Sci. **329:** 364–368.
20. Gilbert, E. & S. Marks. 1979. An analysis of the mortality of workers in a nuclear facility. Radiation Res. **79:** 122–148.
21. Stewart, A., G. Kneale & T. Mancuso. 1980. The Hanford data—a reply to recent criticisms. Ambio **9** (2): 66–73.
22. Linos, A., J. Gray, A. Orvis, R. Kyle, W. O'Fallen & L. Kurland. 1980. Low-dose radiation and leukemia. N. Engl. J. Med. **302** (20): 1101–1105.
23. Gibson, R., S. Graham, A. Lilienfeld, *et al.* 1972. Irradiation in the epidemiology of leukemia among adults. J. Natl. Cancer Inst. **48:** 301–311.
24. Gibson, R., I. D. J. Bross, S. Graham, *et al.* 1968. Leukemia in children exposed to multiple risk factors. N. Engl. J. Med. **279:** 906–909.
25. Boice, J. D. & C. E. Land. 1979. Adult leukemia following diagnostic X-rays? (Review of report by Bross, Ball, and Falen on a tri-state leukemia survey). Am. J. Public Health **69:** 137–145.
26. Johnson, C. J. An evaluation of brain cancer, melanoma, and respiratory cancer of employees of the Rocky Flats Nuclear Weapons Plant in Jefferson County, Colorado. Presented to the Epidemiology Exchange Session at the Annual Meeting of the American Public Health Association, Detroit, Oct. 22, 1980.
27. Center for Disease Control, United States Dept. of Health, Education, and Welfare. Morbidity and Mortality Weekly Report: Leukemia among persons present at an atmospheric nuclear test (Smoky). **28:** 361–362, 1979.
28. Caldwell, G. G., D. B. Kelley, C. W. Heath, Jr. 1980. Leukemia among participants in military maneuvers at a nuclear bomb test. J. Am. Med. Assoc. **244** (14): 1575–1578.

29. NAJARIAN, T. & T. COLTON. 1978. Mortality from leukemia and cancer in shipyard nuclear workers. Lancet **1:** 1018–1020.
30. COLTON, T., R. GREENBERG & J. BARRON. 1980. An analysis of deaths among nuclear shipyard workers. Final Report to National Institute of Occupational Safety and Health, Department of Health and Human Services, June 20.
31. MATANOSKI, G. M. 1981. *In* Cancer: Achievements, Challenges, and Prospects for the 1980s. J. Burchenal & H. Oettgen, Eds. Proceedings of the 1980 International Symposium on Cancer, Sept. 14–18. Grune & Stratton, New York.
32. BERTELL, R. 1977. X-ray exposure and premature aging. J. Surg. Oncol. **9:** (4): 379–391.
33. ROTBLAT, J. 1978. The risks for radiation workers. Bull. At. Sci. **34** (7): 41–46.
34. BARCINSKI, M. A., M. ABREU, J. di ALMEIDA, *et al.* 1975. Cytogenetic investigation in a Brazilian population living in an area of high natural radioactivity. Am. J. Human Genet. **27:** 802–806.
35. KOCHUPILLAI, N., I. C. VERMA, M. S. GREWAL & V. RAMALINGASWAMI. 1976. Down's syndrome and related abnormalities in an area of high background radiation in coastal Kerala. Nature **262:** 60–61.
36. GENTRY, J. T., E. PARKHURST & G. V. BULIN, JR. 1959. An epidemiological study of congenital malformations in New York State. Am. J. Public Health **49:** (4): 497–513.
37. BROSS, I. D. J. & N. NATARAJAN. 1980. Cumulative genetic damage in children exposed to preconception and intrauterine radiation. Invest. Radiol. **15:** 52–64.
38. OPPENHEIM, B. E. 1979. Genetic damage from diagnostic radiation? A critique of the Bross and Natarajan study. J. Am. Med. Assoc. **242** (13): 1390–1393.
39. JONES, P. K. 1980. Critical Review of Bross and Natarajan, 1980. Invest. Radiol. **15:** 65–67.
40. SHIONO, P. H., C. S. CHUNG & N. C. MYRIANTHOPOULOS. 1980. Preconception radiation, intrauterine diagnostic radiation, and childhood neoplasia. J. Natl. Cancer Inst. **65** (4): 681–686.
41. TAMPLIN, A. 1980. Biological effects of radiation (in the low dose range). Information Bulletin, Natural Resources Defense Council, Inc. June 3.
42. ROTBLAT, J. 1977. The puzzle of absent effects. New Sci. **75:** 475–476.
43. MORGAN, K. Z. 1975. Suggested reduction of permissible exposure to plutonium and transuranium elements. Am. Ind. Hyg. Assn. J. **36** (8): 567–575.
44. LYON, M. F., D. G. PAPWORTH & R. J. S. PHILLIPS. 1972. Dose rate and mutation frequency after irradiation of mouse spermatogonia. Nature New Biol. **238:** 101–104.
45. MARSHALL, E. 1981. New A-bomb studies alter radiation estimates. *In* News and Comment. Science **212:** 900–903.

Reappraisal of Potential Risks Associated with Low-Level Radiation[a]

ROSALYN S. YALOW

Solomon A. Berson Research Laboratory
Veterans Administration Medical Center
Bronx, New York 10468

Department of Clinical Sciences
Albert Einstein College of Medicine at
Montefiore Hospital and Medical Center
Bronx, New York 10467

We live in a world in which perception of reality is too often confused with reality, and there are few fields in which more confusion exists than in the popular perception of the hazards of exposure to low-level radiation and low-level radioactive wastes.

Acute exposure to very high radiation doses can kill in days; chronic exposure to more moderate doses also may have fatal consequences. Marie Curie died from aplastic anemia secondary to radiation exposure; so did her daughter, Irene, as did many of her scientific contemporaries. We do not forget the young girls who painted radium dial watches, ingested the radium, and died from bone cancer or the survivors of the bombings of Hiroshima and Nagasaki who subsequently died from leukemia and other radiation-induced cancers. On the other hand, we must also appreciate that from the dawn of mankind our progenitors have been exposed to radiation from the natural radioactivity of soil and food and from extraterrestial cosmic rays. Nonetheless, the path of evolution has been upwards. It is thus evident that exposure to the low-levels of natural radiation has not proven to be deleterious. The question to be addressed in this report is "How low is low"—or are there levels of radiation below which one cannot discern harmful effects.

Most of what is known about the effects of radiation has been obtained from exposure to high radiation doses and dose rates. For a number of years national and international standards-setting bodies concerned with establishing radiation protection guidelines accepted the hypothesis of a linear dose-effect extrapolation with no threshold, on the assumption that this provided a generous safety factor for predicting possible radiation-induced deleterious effects. According to this hypothesis, there would be the same number of radiation-related deaths among a thousand people receiving 100 rem exposure as among a million people receiving 0.1 rem exposure. (A rad is a unit of energy absorbed from ionizing radiation per unit mass and corresponds to 100 ergs/g. A rem includes a factor that takes into account the

[a]**Editor's note:** This presentation was made at the November 4, 1981 meeting of the Section of Science and Public Policy of the New York Academy of Sciences. Dr. Yalow is commenting on the paper presented on November 5, 1980 by Dr. Evelyn Mauss, which appears on pages 27–36.

0077–8923/83/0403–0037 $01.75/2 © 1983, NYAS

different biological effectiveness of lightly and densely ionizing radiation.) Similar considerations have not been promoted for other known carcinogens, such as cigarette smoking. It is rather obvious that at high doses this linear extrapolation represents an over-simplification. For instance, 400 rem acute exposure to 100 people would result in about 50 deaths within a few days or weeks, but 0.4 rem exposure to 100,000 people would certainly not result in 50 radiation-related deaths. Apart from this extreme situation there is an enormous body of evidence from animal studies demonstrating that the effectiveness per unit dose of low-linear energy transfer (LET) radiation (radiation characteristic of electrons, x-rays and gamma rays) for induction of mutagenic alterations or cancer is considerably lower at low-dose rates than at high-dose rates.[1]

In 1980, after much discussion, analysis and controversy the National Academy of Sciences Committee on the Biological Effects of Ionizing Radiation (BEIR III) concluded[2] that the linear extrapolation is unnecessarily conservative and concluded that a linear-quadratic model is more consistent with epidemiologic and radiobiologic data for low-LET radiation. According to this model, radiation effects at low-levels will be *overestimated* by linearly extrapolating from high dose-level responses. It is important to appreciate that extrapolation beyond the range of measurements does not always give true answers even in physics, and physical laws are much simpler than those which govern a complex biologic organism such as man. Thus, repeated confirmation of Newton's Laws of Mechanics, which appeared equally applicable to planetary motion or to a stone dropped from the Leaning Tower of Pisa, gave no insight into the laws governing the behavior of atomic or subatomic particles or the behavior of particles with speeds approaching the velocity of light. If we want to determine the nature of radiation effects at dose rates and doses comparable to those attributable to natural background radiation, it would seem most reasonable to examine populations in regions of the world where background radiation is higher than that to which most of the rest of the world is exposed.

Such a study was performed in China by examining 150,000 Han peasants with essentially the same genetic background and same life style. Half of them lived in a region where they received an almost threefold higher radiation exposure because of radioactive soil.[3] More than 90 percent of the progenitors of the more highly exposed group had lived in the same region for more than 6 generations. The investigation included determination of radiation level by direct dosimetry and evaluation of a number of possible radiation-related health effects including chromosomal aberrations of peripheral lymphocytes, frequencies of hereditary diseases and deformities, frequency of malignancies, growth and development of children and status of spontaneous abortions. This study failed to find any discernible difference between the inhabitants of the two areas. The authors of this study concluded that either there may be a practical threshold for radiation effects or that any effect is so small that the cumulative radiation exposure to three times the usual natural background resulted in no measurable harm after six or more successive generations. Some might say that the Han peasant group studied was resistant to radiation on a genetic basis. If so, could this not be considered to be a radiation-induced enhancement of the quality of the genetic pool?

Let us consider two papers that Dr. Mauss referred to in her presentation to the November 5, 1980 meeting of the Section on Science and Public Policy. One, by Barcinski *et al.*,[4] described a cytogenetic investigation in a Brazilian population living in an area of high-natural radioactivity and the other, by Kochupillai *et al.*,[5] was concerned with Down's syndrome in the high-background radiation area (1.5–3.0 rem/year in contrast to the background of about .1 rem/year in most of the United States) in Kerala, India. While Dr. Mauss suggests that the chromosomal aberrations observed were related to the increased external radiation exposure, the authors concluded that "the high incidence of two-hit type aberrations (were consistent with) the existence of internally deposited sources of high linear energy transfer radiation emitters which could more effectively induce this complex type of aberration." The Brazilian study reported[4] no significant difference between the control and higher radiation exposure groups in number of aneuploid cells, chromatid-type aberrations or chromosome deletions. Whether the other cytogenetic data are significant is open to question since the errors were rather large. For instance, in the higher exposure group the incidence of dicentrics was 0.07 $\pm$ 0.28 (S.D) compared to 0.04 $\pm$ 0.19 in the control group. Some might conclude that the incidence of dicentrics was almost doubled from 0.04 to 0.07. Surely a difference of 0.03 $\pm$ 0.34 can hardly be considered significant. Furthermore, it should also be noted that an earlier study[6] failed to reveal cytogenetic abnormalities in this group.

The data in the Kerala paper[5] must also be examined. The observed incidence of Down's syndrome in the more highly exposed study population of 12,918 was 12 or 1 in 1076. The incidence in the control population of 5938 was 0. This difference appears to be due to statistical variations since an earlier report from the same institution[7] gave the incidence of Down's syndrome in 58,325 births in India as averaging 1 in 1215—not different from the 1 in 1076 in Kerala. It is not without interest that Verma and Singh point out[7] that an earlier World Health Organization study had reported no cases of Down's syndrome among 59,689 births in India although that was contrary to the clinical experience of Indian pediatricians. Verma and Singh[7] referred to other surveys suggesting that the frequency of Down's syndrome throughout the world is quite comparable to that in Kerala. Thus, the absence of Down's syndrome in the control population was an aberration. Dr. Mauss failed to mention that the rat and human populations in Kerala had been well-studied by other investigators[8–10] with respect to the incidence of cytogenetic aberrations and demographic data such as fertility index, sex ratio and infant mortality. In these papers it was concluded that there was no evidence for radiation-related mutational effects.

Let us consider now the regions in the United States where natural background radiation is increased. In Colorado and in Utah the average radiation exposure is about twice that on the East and West Coasts because of increased cosmic rays at their higher elevations and natural radioactivity of the soil. However, death rates due to cancer in these states are among the lowest in the country. It is possible that an appropriate statistical analysis would reveal that the racial, ethnic, age distribution or other factors might account for the low cancer death rates in these two states—after all Mormons neither smoke nor drink. When Mason and Miller[11] compared the age-adjusted risk ratio for mortality from malignancies for Caucasians in Denver and Salt Lake City with those in San Francisco and King, Washington, they observed that the leukemia incidence was slightly but not significantly lower and the incidence of other cancers was significantly lower in the higher radiation exposure cities. An inverse relationship between elevation (hence higher radiation exposure) and mortality from leukemias and lymphomas has also been reported.[12] Others have concluded[13] that in the United States there is no relation between increased background radiation and leukemia.

It is evident from studies such as these that increased radiation exposures to levels as much as three times and perhaps as much as ten times the usual natural background does not result in any measurable increase in deleterious health effects. Effects from man-made radiation at comparable dose-rates and cumulative doses are expected to be similar to those from natural radiation.

Numerous animal studies have demonstrated that excessive radiation at moderate dose-rates leads to a shortening of life primarily due to carcinogenesis. However, it is worthwhile to recall several studies demonstrating *life-lengthening* associated with more modest radiation exposure. Lorenz *et al.*[14] observed that mice exposed to 0.1 rad per day for their entire lives had a mean life expectancy slightly greater than their controls, which were housed separately. Similar findings have been reported by others.[15,16] The biologic mechanisms involved in these unexpected findings is not well understood. One possible explanation is that the improved survival under low-level radiation is not due to prolongation of the natural life span but rather results from a stimulatory response to low-grade injury that produces better health in the exposed group than the unirradiated controls. I am not recommending that we go to spas to drink the radioactive waters as has been done in the past. Nonetheless, these data do demonstrate that radiation exposures considerably in excess of natural background have not proven to be harmful.

There is general agreement that an evaluation of the carcinogenic effects of low doses of low-LET radiation and other weak carcinogens cannot be done directly since the sample size required to estimate the excess risk is prohibitively large when the possible risk is small. Undoubtedly differences in statistical manipulations account for marked discrepancies among different groups examining the carcinogenic effects of radiation and other weak carcinogens. As Land has discussed[17] even assuming a linear extrapolation for radiation effects, a sample size of 100,000,000 would be required to be certain of an increased incidence of breast cancer associated with acutely exposing both breasts of 35-year old women to 1 rem. Such a sample is hardly practical. Therefore, a case-control approach, in which the sample consists of a fixed number of cancer cases and a fixed number of matched non-cases or controls, is used. Land has calculated[17] that using this cohort approach only 1,000,000 women would be require to be certain of a radiation effect from 1 rem. Of course, in the case-control approach to evaluation of radiation and other carcinogens, a sufficient number of subjects are never included and there is not random selection of cases and controls. Hence, the data presented simply do not have statistical significance and subtle sources of bias could well account for purported observed effects. For instance, MacMahon has reported[18] that children born after their mothers had received one to six pelvic radiographs (average dose per radiograph was 1 rad) were 42 percent more likely to die of cancer in the first 10 years of life than were children not irradiated *in utero.* Using the same case-control method of analysis MacMahon *et al.*[19] also reported that drinking 1 to 2 cups of coffee a day introduced a relative risk of 2.6 in developing cancer of the pancreas and further suggested that coffee drinking at his level can account for more than 50 percent of the cases of pancreatic cancer. Since coffee drinking is familiar and radiation is not, most people would discount that his case-control analysis[19] proves that such modest coffee drinking is a risk factor for pancreatic cancer, particularly since the effect did not appear to be dose-related in men—the risk factor was the same, 2.6, whether consumption was 1–2 cups or greater than 5 cups a day. There are other reasons for reluctance in accepting his analysis. For instance, the risk factor was twice as great for ex-smokers compared to current smokers drinking 1 to 2 cups of coffee per day, a rather unlikely finding since it is commonly accepted that smoking is a carcinogen or promoter of other carcinogens. The MacMahon *et al.*[19] report on the association between coffee drinking and cancer of the pancreas is, however, in a sense less flawed than his earlier report[18] on the association between prenatal radiation and early cancer death. In the latter study there was clearly a bias in that no account was taken of the fact that the exposed mothers had medical conditions that prompted the diagnostic x-rays.

It would be impossible to review all the papers in the field of the relationship between prenatal and other diagnostic radiation and malignancy. However, Dr. Mauss in her presentation also quoted out of context the study by Oppenheim *et al.*[20] The conclusions reached by these authors were as follows:

> The evidence for harmful effects of diagnostic x-ray exposure on the human fetus rests primarily on a group of epidemiological studies that demonstrate an increased incidence of these effects among prenatally or preconceptionally irradiated children, as compared with nonirradiated controls. Each of the studies is subject to criticism on the grounds that exposure to diagnostic x-rays occurred only when there was some medical indication for the examination. Selection of subjects on the basis of medical indication may have introduced biases into the various studies, and the possibility exists that the observed results were attributable to these biases rather than to the radiation exposure itself. Comparisons with studies in which radiation exposure occurred on a non-selective basis (from routine pelvimetry examinations or atomic bomb radiation) demonstrate a significant discrepancy in each instance. This casts some doubt on the validity of the studies involving exposure to diagnostic x-rays, and suggests that such exposure may be less harmful than has been claimed.

The negative evidence that the Oppenheim group describes in its paper includes the failure to find radiation-related effects when maternal irradiation was received on a routine basis, rather than for a particular medical condition. It is not without interest that they report a study in which radiation exposure due to *routine* pelvimetry appeared to result in a lower death rate (1.05 percent) before the age of 10 than occurred in the central group (1.42 percent).

Dr. Mauss also reports incorrectly that "the consensus is that a strong association between increased risk of leukemia and prenatal exposure to diagnostic x-rays has been demonstrated." For this conclusion she references a report from an intergovernmental agency task force on the health effects of ionizing radiation.[22] Although this report does include a table dealing with cancers linked to radiation in particular populations, its conclusion about several reports purported to show low-level radiation effects is "none of them is sufficiently extensive, complete or free from methodological complications to provide conclusive information at present on the effects of radiation at low doses." This is not much different from the conclusion reached in the BEIR III report[2] which stated

> Because of large genetic and environmental variables encountered in human populations, it is very difficult to measure any effects that might be produced by low doses of radiation, such as those used in diagnostic radiology. It is therefore not surprising to find conflicting reports on whether the "spontaneous" incidence of malformations or growth retardation is increased as a consequence of such exposure (some authors report negative findings; others positive). At present, it is impossible on the basis of human studies alone to determine with certainty a dose below which teratologic effects in man are not induced by exposure at sensitive stages in development. As discussed above, such thresholds do, however, probably exist, and they may be higher for protracted or fractionated radiation than for acute single exposures.

In particular in dealing with the problem of increased risk of leukemia and prenatal exposure to radiation, the consensus is that in all studies purported to show this effect the selection of patients having x-rays was not on a random basis and therefore there is no way of calculating the risk.[20,21]

Let us consider another case of selective reporting by Dr. Mauss. In referring to the report of Lyon *et al.*[23] excess childhood leukemia in southern Utah as a consequence of nuclear weapons testing she notes that "Lyon's work was analyzed and criticized by the Department of Energy and General Accounting Office statisticians." She then refers to Lyon's responses to those criticisms.[24] The inference would seem to be that the Government, for a variety of political or economic reasons, is involved in another cover-up and that Lyon[24] successfully defended his position. This implication is far from the truth. Let us examine the original data from Lyon *et al.*[23] In FIGURE 1 are shown the mortality rates from leukemia (lower) and for all cancers, including leukemia (upper), for children presumably living in high and low fallout counties in Utah. The 1944–1950 period represents the pre-fallout control low-exposure cohort. The 1951–1958 group was considered to be the high-exposure cohort, that is, those born during the period of maximum above-ground nuclear bomb testing in Nevada. The second low-exposure cohort was the group born after most, but not all, of the above-ground testing had ended.

From a perusal of Lyon's data it could reasonably be concluded that on the average during the entire 30-year period the high fall-out counties might have had a lower incidence of leukemia than the low fall-out counties, but that the uncertainties in the determinations are so large that one cannot reliably conclude whether or not there is a trend. If one considers the sum of childhood malignancies (leukemia plus other cancer deaths) there appears to be a generally downward trend with the drop in the high-fallout counties being somewhat greater than in the low-fallout counties, although if the standard deviations had been included the differences would not have been significant.

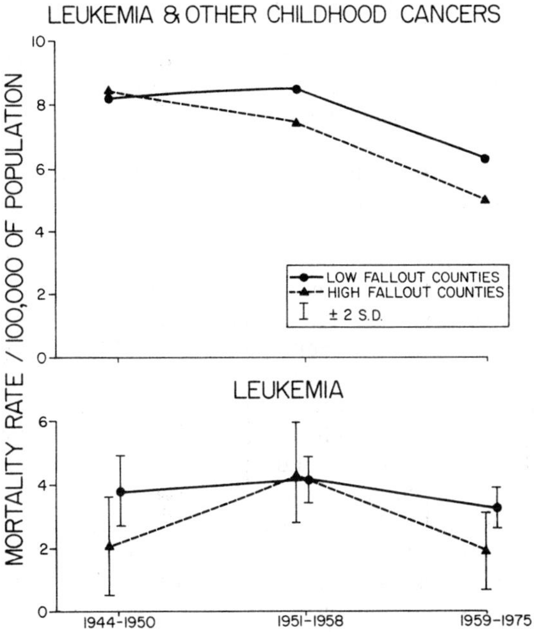

FIGURE 1. Data reproduced in graphical form from reference 22. Total childhood malignancies including leukemia (*top*) and childhood leukemia in high fall-out and low fall-out regions of Utah according to study of Lyon *et al.*

The news headlines[25] following interviews with Dr. Lyon would have been less sensational if he had reported that his data had shown no relation between the totality of childhood malignancies and the atomic tests of the 1950s; rather than selectively reporting what would appear to me, as an experimental scientist whose results do not depend on marginal p-values, to be an inconclusive study on the relation between leukemia and fall-out.

However, his reporting data selectively to the public generated the need for expensive analysis and reanalysis and a little-known scientist acquired a national reputation. Lyon's paper[23] in the *New England Journal of Medicine* was critized in the same and later issues of the *Journal* by others who are more sophisticated in statistics than I. Land,[26] one of our major biostatisticians who, in fact, was quoted favorably by Dr. Mauss, wrote an editorial that also called attention to the unchanged total childhood malignancy rate and reminded us that "It should be remembered, however, that it is in the nature of things for cancer mortality rates based on small populations to vary widely over time." Bader[27] called attention to a review of childhood leukemia in Seattle-King County, Washington, which is not located in an area of fall-out and whose population was slightly larger than the entire state of Utah, which would tend to decrease statistical variation. The number of leukemia cases in successive 5-year periods beginning in 1950 were 37, 50, 56, 44 demonstrating that even in larger populations than the rural districts of Utah "clustering," or statistical variation, is routinely observed. Dr. Bader concluded[27] that "Although tie-ins between cases of several clusters were often established, they were judged to be related to expected statistical probability rather than other facts. . . The caution advised in Land's editorial interpreting the results in Utah seems well advised." In the same issue of the *New England Journal of Medicine,* Enstrom[28] from the University of California School of Public Health, pointed out "In my opinion, the paper by Lyon and his colleagues on deaths from childhood leukemia in Utah represents overinterpretation of ad hoc observational data of limited quality . . . the specific effect, if any, of fallout cannot be determined from the analysis of Lyon and his coworkers."

Enstrom participated in the symposium in which Lyon's "further information" paper was presented and rebuts the purported further information in a paper entitled "The Non-association of Fallout Radiation with Childhood Leukemia." His criticism[29] was based on the inherent errors of small sample statistics, the known effect of clustering, the unknown doses and therefore inability to relate dose to effect and most importantly on the fact that an increased incidence of childhood leukemia was observed in the Japanese children only if their mothers had received doses larger than 100 rads during pregnancy.[30] From the latter study[30] it is known also that leukemia peaked in 1950–52, or 5 to 7 years after the bomb, and decreased thereafter. From Lyon's data on above-ground testing[23] and the year of death for the high exposure cohort that he tabulated for an analysis of radiation effects by the Government Accounting Office,[31] there are some strange inconsistencies (TABLE 1). Thus, of the 32 leukemia deaths recorded for the high exposure cohort, 13 occurred in 1959 and 1960. Allowing for about a 5-year latency period apparently there would appear to be no significant excess mortality from the tests in 1957, the year of the greatest bomb testing. It is indeed rather remarkable that Lyon has not given more publicly to the fact that the clustering in a two-year period, not related to the year of maximum testing, accounts for virtually all of the excess leukemia mortality that he describes in the high exposure cohort. This information, that Dr. Lyon eventually supplied to the GAO, was not included in his so-called rebuttal[24] to the GAO report.[31] I believe that if the reviewers for the *New England Journal of Medicine* had had available the data subsequently provided to the GAO, the paper[23] would not have been accepted. It is not without interest that in the Bader paper,[27] which included a year-by-year listing of leukemia cases in non-irradiated

Seattle-King County, there were only 2 cases in 1959 and there were 20 in 1963 among the 217 cases reported from 1950 to 1972. Thus, a 10-fold difference in annual incidence rates, when the number of cases is small, simply represents statistical variation.

Another report that has received publicity far beyond its scientific merit is the Mancuso study of workers at the Hanford Laboratories, the site of several of the then AEC's reactors. The history of this study dates back to 1964 when Dr. Mancuso was awarded a contract to investigate for the AEC the health of these workers. Dr. Barkev Sanders, a statistician, and Dr. Allen Brodsky, a

TABLE 1. Year of Bomb Testing in Nevada[a] and Year of Death From Leukemia for the Southern Utah High-Exposure Cohort[b]

	Test Yield (kilotons)	Number of Deaths Southern Utah High Exposure Cohort
51	112	0
52	104	3
53	252	2
54	—	0
55	167	2
56	?	2
57	343	0
58	38	0
59	—	7
60	—	6
61	—	2
62	101	1
63	<20	1
64	<20	0
65	<80	1
66	<60	1
67	20–200	2
68	<60	1
69	20–200	1

[a]Data taken from reference 22.
[b]From data supplied by Dr. Lyon for the report in reference 31, pages XVIII–43.

health physicist, were co-investigators in this project. Annual project reports for many years suggested only that there were negative findings regarding a link between cancer and radiation and Drs. Sanders and Brodsky left the project. There were no papers published during the period in which they were involved in the analysis of the data. In about 1976, Mancuso was joined by the pediatrician Stewart and the statistician Kneale who had acquired a reputation for their studies on relationships between diagnostic x-rays and childhood cancers. Together they wrote a controversial paper[32] purporting to show that workers at Hanford had a statistically significant increase in the incidence in two types of cancer, multiple myeloma and cancer of the pancreas. According to Table 10 of this paper,[32] the mean cumulative radiation dose for Hanford workers who subsequently died from cardiovascular disease was 1.05 rads; for solid tumors 1.3 rads; for leukemias and lymphomas 2.2 rads. This excess radiation exposure is quite comparable to that received in a 10 to 20 year residence in Colorado—with its low cancer death rate. The statement of the Mancuso report[32] that has not been widely publicized was that Hanford workers receiving the highest radiation doses (greater than 15 rem) had a lower death rate from all causes and from all malignant neoplasms than expected in a control population. However, because of the small numbers of workers who received this exposure and the small number of cancer deaths in this group, (a total of 14 cancer deaths) the distribution among the different malignancies appeared to have a pattern not identical with that found in much larger groups. Subsequently, both Dr. Brodsky and Dr. Sanders, who initially collaborated with Dr. Mancuso, have been highly critical of the Mancuso paper. An independent analysis by Gilbert and Marks[33] is most revealing. The positive correlation purported to be demonstrated in the Mancuso report appears to be due to 3 deaths from pancreatic cancer in workers receiving more than 15 rem cumulative exposure. According to Gilbert and Marks[33] this diagnosis had been confirmed only in 1 case. If one were to accept the MacMahon report[19] linking coffee drinking with cancer of the pancreas, perhaps a history of coffee habits rather than radiation exposure would be relevant! Furthermore, it should also be noted that the Atomic Bomb Casualty Committee report[35] described a much larger cohort with much higher radiation exposures (up to 400 rems) and found no positive link between pancreatic neoplasms and radiation.

The second category of excess cancer deaths was reported to be multiple myeloma, which included 3 cases compared to an expected number of 0.6. Whether this excess of 2 deaths in this category represents a statistical variation or the effect of another carcinogen cannot be determined. Nonetheless, since among those receiving a cumulative exposure of 15 rems, the observed number of subjects with malignancies was only 14 compared to an expected 24 in a control population, could one not be tempted to conclude, in keeping with the earlier studies I described, that radiation at this level is protective against malignancies?

It would be possible for me to continue to evaluate the other studies to which Dr. Mauss refers. However, most of these have been *considered* in great detail and *refuted* by a committee of experts, a panel of the National Committee on Radiation Protection.[1]

It is perhaps worth mentioning a study of the largest group, apart from the Japanese bomb survivors, who were exposed rather acutely to generalized body radiation significantly greater than that from natural background.[35] This group consisted of 36,000 hyperthyroid patients from 26 medical centers, approximately 22,000 of whom were treated with radioactive ^{131}I and most of the remainder with surgery. The average bone-marrow blood dose due to the ^{131}I was about 8 rads, half of which is delivered within one week; the thyroidal dose generally exceeded 10,000 rems. The follow-up for the ^{131}I-treated group averaged 7 years, quite comparable to the delay time of peaking of leukemia incidence (5 to 7 years) following atom bomb exposure. The conclusion reached[35] was that the incidence of leukemia does not differ between

^{131}I-treated and surgically treated hyperthyroid subjects. This study then is consistent with the inability to demonstrate induction of malignancy due to an acute radiation dose in the 5 to 10 rem range.

Perhaps it is worthwhile calling attention to another large group receiving significant radiation exposure. Airplane crews and passengers receive exposures of 0.5 to 1 mrem per hour when flying at 35,000 feet, due to the absorption of increased cosmic radiation. Workers at highest risk appear to be airline crews who service transcontinental or transoceanic flights because of the longer flying times at greater heights. I am considered a radiation worker since my professional work involves the application of radioisotopes in biomedical investigation and clinical medicine. Yet my radiation exposure in the laboratory is low compared to that which I receive while traveling. Generally, I do not wear a radiation badge when I fly. However, on a recent trip to South America I carried a calibrated radiation monitor and measured my exposure. It was 10 mrem, more than I have ever received in one week in my laboratory. The hazards of a successful radiation worker seem to be that one receives more radiation exposure in travel than in the laboratory! Public concerns have been generated by television and newspaper coverage about radiation received by workers in nuclear plants or even the population living nearby. Yet there is no media communication to the public about the radiation exposures associated with modern air travel. Radiation exposures of those working with radioactive materials or x-radiation are monitored and recorded but not those of air line crews who on the average receive comparable radiation doses.[2]

It is of interest to compare the total exposure person-rems of different categories of people since if there were no threshold for radiation induction of cancer and the linear hypothesis held, that total cumulative exposure of all persons would determine the total toxic effect. According to the latest BEIR report,[2] in 1973 airline passengers and crew accumulated 105,000 person-rem and workers in commercial power reactor plants accumulated 14,000 person-rem. The best estimate of the population dose resulting from the Three Mile Island nuclear accident was 3300 person-rem. It would seem from data such as these that the airline industry causes significantly more radiation exposure to the population than does the nuclear power industry—but the public is never informed about the relative radiation exposure as a consequence of these two industries. I am not suggesting that the radiation exposure of airline crews represents a health hazard. After all, their average cumulative yearly exposure is no more than the increased radiation exposure received by those millions who live in Colorado rather than in New York. Surely one should not recommend radiation monitoring of those who live on the Colorado plateau.

The last scientific topic I will consider is the risk of genetic effects from radiation. This was considered extensively in the BEIR III report[2] and it was concluded that since radiation-induced transmitted genetic effects have not been demonstrated in man, estimates of genetic risk must be based on laboratory data at high dose rates. Recently, in *Science,* Schull *et al.*[36] concluded on the basis of studies of the children born to survivors of the Hiroshima-Nagasaki bombings, that the estimated doubling dose for genetic changes would be about 150 rem, a value some fourfold higher than the results from experimental studies on mice. Furthermore, this represents simply an estimate since they reported that in *no* instance was there a statistically significant effect of parental exposure. It should be noted also that many investigators have found that chronic irradiation in mice is about threefold less effective than acute irradiation. This would effectively raise the doubling dose in man to about 500 rem. Furthermore, since none of the studies in mice were performed at truly low dose rates, one cannot really determine what the doubling rate would be for background radiation.

In my introduction to this presentation I raised the question "How low is low?" or are there levels of radiation below which one cannot discern deleterious effects of radiation. The answer to that question is yes. The GAO report concludes[31] that "there is as yet no way to determine precisely the cancer risks of low-level ionizing radiation exposure, and it is unlikely that this question will be resolved soon." Stating that there are levels below which one cannot discern harmful effects of radiation is not the same as reaching conclusions concerning the existence of a threshold below which radiation effects in man cannot occur. In science we can only accept as valid those laws that are subject to experimental proof. We can hypothesize, but we should not confuse hypothesis with reality. I agree with the GAO report conclusion concerning radiation studies that "the uncertainties in this area pose particular problems for Federal agencies in deciding how best to spend limited research dollars." I do not agree that continued Federal support of research in this area is cost-effective or desirable. After 80 years, 80,000 papers and an investment of about two billion dollars, it is evident that epidemiologic studies cannot produce meaningful data about the existence of a threshold for radiation effects. Molecular and cellular studies may or may not give some insight about molecular or cellular effects but cannot answer important questions about repair mechanisms in the intact animal or man when radiation is delivered at low dose-rates. At present there are no really good ideas that would permit a breakthrough in the field of low-level radiation effects. Therefore, good scientists with imagination and insight are unlikely to work in a field in which the studies, at best, are pedantic and, at worst, are inevitably flawed. Thus, this field of research, which now primarily exists through Federal concern and congressional mandate rather than scientific interest, is most likely to attract mediocre investigators and result in inconclusive studies. Dr. Mauss concluded that "The studies on which conclusions (about low-level radiation effects) must be based are for the most part disparate not prospective, and not replicable." She might have added that studies that are disparate and not replicable are simply not contributions to science.

What are society's rewards for investment in research? The ultimate reward is that the knowledge gained can be used in the service of man. Mancuso's inconclusive study of the Hanford workers cost $6 million. In contrast, three decades of my own work, which has had important effects on human health and has had relevance in tens of thousands of investigations in medicine and biology, cost the government less than a third as much. In conclusion, it is essential to communicate to the public and through them to our government that each of us loses when scientific talent and funding is diverted from scientifically important and socially desirable investigations to predictably negative experimentation because of irrational fears generated by well-intentioned or ill-intentioned but often uniformed Cassandras.

Let me turn now, for a moment, from science to philosophy. Science, unlike religion, is not influenced by belief or divine revelation. In science we observe, hypothesize and reobserve in an attempt to determine whether the hypothesis is consistent with the observations. Scientists are not Aristotelians—if we want to determine the number of teeth in a horse's mouth we open its mouth and count them. Sometimes the critical experiments cannot be performed with available tools. Thus, Newton's Laws, which were dogma for over two centuries, could not predict behavior at high velocities or atomic or subatomic dimensions. But this is simple compared to predicting the laws governing the much more complex interrelations in biological systems including man. With respect to low-level radiation effects, there is a consensus that there are no reproducible studies demonstrating unequivocally that such effects are demonstrable. The disagreement concerns how to extrapolate from higher dose rates and total doses to the non-measurable range. The BEIR III report concluded that a linear-quadratic extrapolation is the one most consistent with available data; Radford dissented and upheld that the linear extrapolation hypothesis was appropriately conservative; Rossi claimed that a quadratic extrapolation should be employed. There are those who hold that there is not solid evidence excluding the possibility that there is a threshold below which there are no radiation effects. After all, with what we have learned in recent years through studies in molecular biology there is every reason to believe we have repair mechanisms hitherto undreamed of. The disagreement in the low-level radiation field is about hypothesis—not observable facts. One could not determine the validity of Newton's Laws at subatomic dimensions until the tools become available. However, in that case there was no need to make policy decisions based on extrapolation. In the case of low-level radiation effects, public policy decisions need to be made in the absence of scientific evidence. It should be appreciated that these are arbitrary decisions based on philosophy not fact and may well change because of political or other considerations. An interesting conflict-of-interest does arise when those recommending additional funding be provided for investigations in this area are the ones most likely to profit thereby. My philosophical approach is that funding be withdrawn from research in this field until better tools become available—if they ever do—and that a cost-benefit analysis be required for any new regulatory decisions. There is a distinction between conjecture and scientific fact, and that distinction has not been made by those concerned with potential risks associated with low-level radiation.

REFERENCES

1. National Council on Radiation Protection and Measurements. 1980. Influence of Dose and its Distribution in Time on Dose-response relationships for Low-LET Radiations. NCRP, Report No. 64. Washington, D.C., April.

2. NATIONAL ACADEMY OF SCIENCES, COMMITTEE ON THE BIOLOGICAL EFFECTS OF IONIZING RADIATIONS, ASSEMBLY OF LIFE SCIENCES. 1980. The Effects on Populations of Exposure to Low-Levels of Ionizing Radiation, (BEIR III Report) NAS/NCRP Washington, D. C.
3. HIGH BACKGROUND RADIATION RESEARCH GROUP, CHINA. 1980. Health survey in high background radiation areas in China. Science **209:** 877–880.
4. BARCINSKI, M. A., M.-D., A. ABREU, J. C. C. DE ALMEIDA, J. M. NAYA, L. G. FONSECA & L. E. CASTRO. 1975. Cytogenetic investigation in a Brazilian population living in an area of high natural radioactivity. Am J. Hum. Genet. **27:** 802–806.
5. KOCHUPILLAI, N., I. C. VERMAN, M. S. GREWAL & V. RAMALINGASWAMI. 1976. Down's syndrome and related abnormalities in an area of high background radiation in coastal Kerala. Nature **262:** 60–61.
6. CULLEN, T. L. 1970. Dosimetric and Cytogenetic Studies in Brazilian Areas of High Natural Activity. Health Phys. **19:** 165–166.
7. VERMA, I. C. & M. SINGH. Down's Syndrome in India. Lancet **i:** 1200. 1975.
8. GRUNEBERG, H., G. S. BAINS, R. J. BERRY, L. RILES, C. A. B. SMITH & R. A. WEISS. 1966. A Search for Genetic Effects of High Natural Radioactivity in South India. Med. Res. Council Special Report Series No. 307, H.M.S.O. London.
9. GOPAL-AYENGAR, A. R., K. SUNDARAM, K. B. MISTRY, C. M. SUNTA, K. S. V. NAMBI, S. P. KATHURIA, A. S. BASU & M. DAVID. 1972. Evaluation of the long-term effects of high background radiation on selected population groups on the Kerala Coast. Peaceful Uses of Atomic Energy **II:** 31–51.
10. AHUJA, Y. R., A. SHARMA, K. U. K. NAMPOOTHIRI, M. R. AHUJA & E. R. DEMPSTER. 1973. Evaluation of effects of high natural background radiation on some genetic traits in the inhabitants of monazite belt in Kerala, India. Hum. Biol. **45:** 167–179.
11. MASON, T. J. & R. W. MILLER. 1974. Cosmic radiation at high altitudes and U.S. cancer mortality, 1950–1969. Radiat. Res. **60:** 302–306.
12. GRAIG, L. & H. SEIDMAN. 1960. Leukemia and lymphoma mortality in relation to cosmic radiation. Blood **17:** 319–327.
13. JACOBSON, A. P., F. A. PLATO & N. A. FRIGERIO. 1976. The role of natural radiations in human leukemogenesis. Am. J. Public Health **66:** 31–37.
14. LORENZ, E., J. W. HOLLCROFT, E. MILLER, C. C. CONGDON & R. SCHWEISTHAL. 1955. Long-term effects of acute and chronic irradiation in mice. I. Survival and tumor incidence following chronic irradiation of 0.11 R per day. J. Natl. Cancer Inst. **15:** 1049–1058.
15. CARLSON, L. D., W. J. SCHEYER & B. H. JACKSON. 1957. The combined effects of ionizing radiation and low temperature on the metabolism, longevity, and soft tissues of the white rat. I. Metabolism and Longevity. Radiat. Res. **7:** 190–197.
16. SACHER, G. A. & D. GRAHN. 1964. Survival of mice under duration-of-life exposure to gamma rays, I. The dosage-survival relation and the lethality function. J. Natl. Cancer Inst. **32:** 277–321.
17. LAND, C. E. Estimating cancer risks from low doses of ionizing radiation. 1980. Science **209:** 1197–1203.
18. MACMAHON, B. 1962. Prenatal x-ray exposure and childhood cancer. J. Natl. Cancer Inst. **28:** 1173–1191.
19. MACMAHON, B., S. YEN, D. TRICHOPOULOS, K. WARREN & G. NARDI. 1981. Coffee and cancer of the pancreas. New Engl. J. Med. **304:** 630–633.

20. OPPENHEIM, B. E., M. L. GRIEM & P. MEIER. 1975. The effects of diagnostic x-ray exposure on the human fetus: An examination of the evidence. Diagn. Radiol. **114:** 529–534.
21. TOTTER, J. R. & H. G. MACPHERSON. 1981. Do childhood cancers result from prenatal x-rays? Health Phys. **40:** 511–524.
22. INTERAGENCY TASK FORCE ON THE HEALTH EFFECTS OF IONIZING RADIATION. 1979. Report of the Work Group on Science. U.S. Dept. of Health, Education, and Welfare, Washington, D.C.
23. LYON, J. L., M. R. KLAUBER, J. W. GARDNER & K. S. UDALL. 1979. Childhood leukemias associated with fallout from nuclear testing. New Engl. J. Med. **300:** 397–402.
24. LYON, J. L., J. W. GARDNER, D. W. WEST & L. SCHUSSMAN. 1980. Further information on the association of childhood leukemias with atomic fallout. *In* Banbury Report No. 4, Cancer Incidence in Defined Populations. J. Cairns, J. L. Lyon & M. Skolnick, Eds. Cold Spring Harbor, NY, pp. 145–162.
25. LYON, J. L. 1980. MD's Study Shows Link Between Leukemia Deaths and Atomic Tests of 1950's. American Medical News, page 18. February 15.
26. LAND, C. E. 1979. The hazards of fallout or of epidemiologic research? New Engl. J. Med. **300:** 431–432.
27. BADER, M. 1979. Leukemia from atomic fallout. New Engl. J. Med. **26:** 1491.
28. ENSTROM, J. E. 1979. Editorial. New Engl. J. Med. **26:** 1491.
29. ENSTROM, J. E. 1980. The nonassociation of fallout radiation with childhood leukemia in Utah. *In* Banbury Report No. 4, Cancer Incidence in Defined Populations. J. Cairns, J. L. Lyon & M. Skolnick, Eds. Cold Spring Harbor, NY, pp. 163–186.
30. JABLON, S., J. L. BELSKY, K. TACHIKAWA & A. STEER. 1971. Cancer in Japanese exposed as children to atomic bombs. Lancet **i:** 927–931.
31. Report to the United States Congress by the Comptroller General. 1981. Problems in Assessing the Cancer Risks of Low-Level Ionizing Radiation Exposure. Vol. 2. U.S. General Accounting Office. Gaithersburg, MD.
32. MANCUSO, T. F., A. STEWART & G. KENEALE. 1977. Radiation exposure of Hanford workers dying from cancer and other causes. Health Phys. **33:** 369–385.
33. GILBERT, E. S. & S. MARKS. 1979. An analysis of the mortality of workers in a nuclear facility. Radiat. Res. **79:** 122–148.
34. BEEBE, G. W., H. KATO & C. E. LAND. 1978. Studies of the mortality of A-bomb survivors, 6. Mortality and radiation dose, 1950–1974. Radiat. Res. **75:** 138–201.
35. SAENGER, E. L., G. E. THOMAS & E. A. TOMPKINS. 1968. Incidence of leukemia following treatment of hyperthyroidism. J. Am. Med. Assoc. **205:** 147–154.
36. SCHULL, W. J., M. OTAKE & J. V. NEEL. 1981. Genetic effects of the atomic bombs: A reappraisal. Science **213:** 1220–1227.

GENERAL DISCUSSION

Moderator: Frances S. Sterrett

E. MAUSS: I want to join Frances Sterrett in thanking Dr. Yalow for this presentation.

A first important point: the fact that, as Dr. Yalow has said, there is no conclusive evidence of a linear dose-response relationship, i.e. no conclusive evidence of a threshhold below which there may be no effect, does not make less useful those decisions that have been made on the basis of a conservative effort to use worst possible cases when thinking of public policy. It seems judicious and reasonable for our public agencies to have assumed a linear hypothesis.

Now on the linear hypothesis, I would like to remind you that a reevaluation of the data on radiation from the Hiroshima, Nagasaki bombs is now in progress. The government is on the verge of spending a great deal of money in the next couple of years to reevaluate as carefully as possible the radiation those bombs produced. And it seems very likely, at least according to the News and Comment articles in *Science* on May 22 and October 2, that the increase in leukemia may be described as a linear quadratic [curve]. (**Note:** Some of this took place on the blackboard.)

R. YALOW: What can I say—the linear extrapolation hypothesis has been shown not to be valid for lightly ionizing radiation.

You can pay your money and take your choice. So they are going to spend an enormous amount of money to continue to recalculate and reexamine the possible doses that would have been received. Remember, we are projecting back 36 years and trying to determine what the neutron component was. Nonetheless, there were enormous variations in exposure due to shielding, etc. From the Nagasaki data, it would appear as if there is no increase in leukemia until doses are quite high, greater than 20–50 rem.

MAUSS: Well, this isn't "nothing." New data just published in *Physics Today* (in October), suggest that instead of being linear, so far from there being no effect at lowest doses, the curve curves up a little bit. Even Charles Land, the National Cancer Institute statistician, has said that on the basis of the new data, it is likely that the risk estimates may have to be doubled. And then if one considers the shielding from buildings and so on, the risk estimates may have to be considerably more than doubled. Whether this means anything or not, whether we want to contrast the deleterious health effects of low-level ionizing radiation with nuclear war effects or traffic accidents or whatever, is a judgment each of us is entitled to make. But if we are looking at the mass of data, those 80,000 papers, many are at the very least highly suggestive of some effects. If we understand that deleterious health effects at low levels exist, then the judgment can be made as to what to do about them.

Part of the question is, do we really want to know? If we do not really want to know because it costs too much, that is a value judgment, to be sure.

Oppenheim, as I said originally, reviewed much of the published material on prenatal and carcinogenic effects of exposure to low-level radiation and found biases in design and biases in the observations.

On the other hand, there are seemingly bias-free observations, and one government agency was unequivocal on the question of leukemia following prenatal exposure. The Inter-Agency Task Force of the then U.S. Department of Health, Education and Welfare reported in 1979 on the health effects of ionizing radiation and indicated that a relationship has been observed. On balance it seems to me rash to say that there are no deleterious health effects.

YALOW: No observable health effects. It is a different thing.

MAUSS: Okay, I do not think we can resolve this right now, and I do want to give other people a chance.

YALOW: I think that we have to put these things in a proper dollar framework. I pointed out that the Mancuso study cost $6 million over a 10-year period; the total support for my research was $2 million over a 30-year period. The screening programs for neonatal hypothyroidism that save one in 4,000 children from irreversible mental retardation cost about $6 million each year. Now you have to examine what you are getting for it. The Land analysis tells us that it is going to take 100 million people to show an effect from one rem on the incidence of breast cancer. There is no controlled study that will permit you to determine such an effect. So that on the average, you will find positive results and you will find negative results. And generally in science, when we find positive results and negative results, we average them out to zero. Except that there is a prejudice in this. Most people who find a positive result report it. And most people who find a zero report or negative report do not report it. So that the literature is now built with people reporting on their prejudices. Really the only mass study that is in a sense reasonable is of the Mormons. It is reasonable because they have the same lifestyle and they live in the same place where there is double the usual radiation exposure. There are now probably 700,000 of them in Utah. And they have lower death rates from cancer and a variety of other things than anybody else. Now this is a good controlled in-bred population. Where are we ever going to get numbers like that again?

UNIDENTIFIED SPEAKER: I would like to ask a question. I attended a symposium the other day and there was great controversy over one question. Perhaps you with your expertise can give us the answer. This controversy had to do with the Hiroshima and Nagasaki bombing. In the area where they had the bombing they are growing a special food that the Japanese export to this country. I personally like that food. I would like to eat it. But I brought up the question inasmuch as it is grown in the soil in the exact same place as the bombs fell. Can the radiation be damaging when you ingest the food?

YALOW: All food is radioactive. I do not know about the increase, if any, in the radioactivity of that food.

UNIDENTIFIED SPEAKER: It is just grains—health food. But it is grown on the same ground.

YALOW: Let me just explain a couple of things to you. Because of the possible uptake of radioactive iodine in the thyroid as a result of fallout from the atom bomb, they instituted a program of examining the thyroids. And they found a very high incidence of carcinoma in situ in the people who lived in Hiroshima and Nagasaki. This looked interesting. It was something like 20% in people over 20.

UNIDENTIFIED SPEAKER: How many years after the bombing?

YALOW: It started in the 1950s. Then they looked at Japanese living in different regions and made the same observations. They examined the Japanese who live in Hawaii and they discovered that it is the same. In our country, we have about a third the Japanese rate of thyroid carcinoma in situ. In Switzerland, they have about 10% of the Japanese rate. Now do you know where people are dying from thyroid cancer? In Switzerland. In Japan, it has a very low incidence.

UNIDENTIFIED SPEAKER: What are you getting at from the question I asked?

YALOW: I am trying to bring out that you have to examine the way in which you look at these things. The initial reports of thyroid cancer in Hiroshima and Nagasaki might have looked terrible until you started examining the appropriate population. I can read to you a report the government just had prepared. You know the Veterans Administration has been plagued by the soldiers who claim they went into Hiroshima or Nagasaki six weeks after the bombings and are dying from cancer of the lung. These soldiers had less exposure than if they had lived 1 year in Denver. But I absolutely assure you that I do not know what carcinogens there are in the Japanese soil, in Japanese food. They have the highest incidence of gastric carcinoma in the world. They had it before there was bombing. If you want to eat what the Japanese eat, you may get gastric carcinoma. So I am not willing to guarantee there are no carcinogens. I am only willing to guarantee that there are no residual radiation effects.

UNIDENTIFIED SPEAKER: I will just give you an example of what I am trying to bring out. I had radiation therapy on my face when I was 16. That is many many years ago. Okay. I went to a doctor and he said that I have to be careful of thyroid cancer because of that radiation.

YALOW: I can give an answer to that one also. When Royce examined a group of patients who had received thyroid radiation, and had them come with a group of control subjects of the same sex and about the same age, siblings or otherwise, and the physicians who examined them did not know who had and who had not had radiation, they found no difference in incidence of thyroid nodules or thyroid cancer. In all the studies that have been reported the physician knew who had radiation, and if you have 14% incidence of thyroid carcinoma in situ and you're worried and you operate, you are going to find it. If you do not operate, you are not going to find it.

L. SOLON (*Bureau for Radiation Control, New York City Health Department*): On June the 4th and 5th of this year, at the New York Society of Ethical Culture Symposium, dealing with the ethical and public policy aspects of the nuclear fuel cycle, one of the participants in the preparation of the symposium was Philip Kotlar, whose memory we honor tonight. I remember that Philip Kotlar was not an experimental scientist or regulatory scientist. He was a teacher and a reporter. I remember vividly in the year and a half or two years that he participated in our deliberations he contributed to the discussions with sensitivity and perception about all the concerns which have been so merrily and so callously and I must say sort of humorously dismissed by our distinguished lecturer tonight. What is the conclusion to be inferred from Dr. Yalow's remarks? That it was a mistake for the Soviet Union and the United States to discontinue their atmospheric weapons tests? And I can tell you, from somebody who was in the 1955 teapot series, that it is a lot more expensive to do "cost effectiveness." It is a lot more expensive to detonate nuclear weapons underground and do your diagnostics of the yields of the weapons than to do atmospheric weapons tests. Are we to conclude from the remarks Dr. Yalow has made about "cost-benefit" that in fact we should resume the atmospheric detonation of nuclear weapons because after all, look at the minimal and insignificant kind of leukemia increase we have experienced? When I started in this business, roughly contemporaneously with Dr. Yalow, about 30 or so years ago, the permissible level for occupational workers like Dr. Yalow and myself (I happened to be Assistant Chief and later Chief of the Health and Safety Laboratory Radiation Branch of the U.S. Atomic Energy Commission) was three-tenths of a rem per week, on the order of 15 rem per year. As those of you who are occupational workers, including Dr. Yalow, know, the occupational exposure permitted now is 5 rem a year. Now is this the result of some kind of propaganda, some kind of misconception, some kind of lack of careful evaluation? Obviously it is much less expensive to accept 15 rem rather than 5 per year. The dismissal of the vast burden of 80,000 papers, $2 billion and 80 years of study, which in my judgment has been so cavalierly performed by Dr. Yalow, is not warranted by the scientific information. Again citing the BEIR III report, which I in no sense regard as Torah or holy writ, but rather as something that certainly has a great deal of scientific information. Although there was polarization between Rossi with his quadratic hypothesis and Radford with the linear hypothesis, there was no indication or implication that one could go down to very low levels of radiation and receive no radiation effects.

I suggest that there be an intensive reexamination of low-level ionizing radiation that is indeed cost-effective. If one subscribes to the notion that Dr. Yalow has advanced to this group tonight, it means a lower focus on the nuclear fuel cycle. It means a lower focus on reactor operations. It means a lower focus on nuclear weapon detonations and developments. And—I hesitate to say it because I do not think it is funny—it implies a certain lower perception and scrutiny as to the meaning of nuclear war. On the weapons test fall-out aspect , as far distant and remote as it is, it remains a serious matter.

I suggest that we listen to Dr. Yalow because she has important things to say, although personally I would rather hear her stick to her vocation of nuclear medicine than her advocacy and personal feelings and perceptions. By and large, I think that the study she has made is not beneficial in terms of public health. It is not beneficial in terms of preparation for augmenting nuclear power. And it is certainly not beneficial in terms of the proliferation question. France and China still, as you know, resort to atmospheric testing.

YALOW: Not Russia? They are doing underground testing.

SOLON: Russia I did not say. I said France and China are doing atmospheric testing. If you subscribe to the hypothesis that has been advanced by Dr. Yalow we should go back to atmospheric testing because it is really cheaper. The United States and the Soviet Union can advance the same arguments to more competently destroy all humanity.

I do remember Philip Kotlar. I remember his sensitivity and his concerns. I respect Dr. Yalow for her contributions to radioimmunoassay and what it means in terms of medicine. But take a very careful look at what she says and what it implies in terms of application to low-level ionizing radiations.

YALOW: That was a very rapid escalation from biologic effects of low-level radiation, to nuclear power, to nuclear war. It is true that the Physicians for Social Responsibility have campaigned actively against nuclear power as well as against nuclear war. I would assume that all rational people oppose nuclear war. However, I believe overdependence on foreign oil is dangerous. Carter's sending ships to the Indian Ocean, presumably to defend against an oil cut-off, is frightening. There is good reason to believe the Japanese attacked Pearl Harbor because their lifeline to oil would be cut off if they were not free to conquer Southeast Asia. I do not want to fight for oil. I believe it is essential to have adequate power—according to the NAS report on Energy in Transition the principal options are coal and nuclear power and there is every reason to believe nuclear power is safer for the wokers involved as well as for the general population.

UNIDENTIFIED SPEAKER: I think that there are an awful lot of things to be afraid of, but it seems to me that if there is a large investigation on the effects of low-level radiation, it does not necessarily mean that we do not use nuclear energy nor does it necessarily mean that they find that low-level radiation is indeed carcinogenic or seriously so. But it seems to me that we are in effect making a large-scale effort to obtain the truth on a subject that is of major concern to new generations, and it also seems to me that your argument is a little like the argument that used to take place between the pro-polio researchers versus the people who felt that the money should go into research on rheumatic fever. I would think it would be better to keep the focus and perhaps lay to rest forever the question of the negative effects of low-level radiation. Or indeed find out that it does have. . . .

YALOW: You see, inherent in the beast is that you simply would have to deal with such a large, carefully controlled human population. This is why the General Accounting Office concluded that you have to go back to cellular mechanisms. One can show that radiation produces chemical changes. But you cannot go from that and prove that there will be changes in complete animals that have a variety of repair mechanisms. I am perfectly willing to accept any investigation in which the plan of the investigation suggests a different approach than what has been done up to now and has been shown not to work. All I have said is let's have a moratorium until somebody comes up with a good idea. And in a sense, this is exactly what happened with polio. Because the original work of the March of Dimes was to make better respirators, to make better things for the treatment of polio. It was the breakthrough of being able to do tissue culture studies and show that you could grow the virus that make the breakthrough. When and if we have the breakthrough with a new idea. . . .

UNIDENTIFIED SPEAKER: But they kept the money coming.

YALOW: Most of the money went to something wasted like the iron lungs.

UNIDENTIFIED SPEAKER: Dr. Yalow, would you care to comment on the idea of doing x-rays and mammographs in women with breast cancer? The reason I am asking this question is that there is quite a pool of fear about the diagnostic use of x-rays.

YALOW: I do not want to give a definitive opinion, because I think you also have to do cost effectiveness. If you are going to survey everybody, what is the probability of picking up a small tumor that you would not otherwise have detected? I think the very critical issue that really has not been addressed is that breast cancer, like many types of cancer, runs in families. So that if your sister has had breast cancer, there is a very high probability that you will have it. Then you get into a very interesting ethical question that radiation may be a co-carcinogen. If you are radiated to look for breast cancer every year, and you are in the group that has the probability of having breast cancer, it might enhance it. On the other hand, it might pick it up at an earlier and treatable stage. If you are not in the group that is likely to have breast cancer, I would not be terribly worried about radiation that you get from mammography nowadays. I just do not think that it is cost-effective for regular mass screening. However, I would not refuse to have it.

UNIDENTIFIED SPEAKER: I have a very short question preceded by a short statement. Actually, I came in here knowing something about Dr. Yalow's work, having been on the panel debate on nuclear energy with her a couple of years ago. But anyone who did not know probably would have assumed that she was giving some kind of objective impartial look at the controversial and uncertain aspects of low-level radiation and interpretation of their effect. But I think what Dr. Yalow just said in her outburst about nuclear energy and war and the Soviet and oil, could be understood in the context that Dr. Solon put his finger on very well: Dr. Yalow did not come in here as a disinterested party, but as a participant in a long, historic campaign in which this is just a step along the way, deeper then just the nuclear controversy, which has do do with the question of what makes science policy in this country. I find it a perfect illustration of why the credibility of scientists is so low today. This concerns not only her accusations that Dr. Mauss clipped her data, which is precisely what the pro-nuclear establishment has done all along, but also saying that her studies are worth more than Dr. Mancuso's. We know that his funding was terminated when he started to find positive correlations. But it is a very interesting illustration of a scientist trying to be at the same time both scientific expert and a promoter. I do have a very simple question. You do not have to understand the numbers. Given the uncertainties and the controversies what is your role? What are the implications for public health in your role?

YALOW: The implications for public health at levels of radiation. . . . ?

UNIDENTIFIED SPEAKER: No, no, the implications if you are wrong about the effect of low-level radiation.

YALOW: The implications for public health if you assume the linear extrapolation hypothesis would indicate that you could not measure the effects. If I am wrong that there is. . . .

UNIDENTIFIED SPEAKER: Maybe you are deliberately misunderstanding.

YALOW: I am deliberately understanding you.

UNIDENTIFIED SPEAKER: But if the effects of low-level radiation are far more serious than you say or than we even know now, what are the implications for the operation of the nuclear power industry and nuclear weapons manufacture? What are the public health implications if you're wrong?

YALOW: As far as I am concerned, it depends on what you mean by "if I am wrong." When we calculated the total amount of tritium and carbon to be disposed of in liquid scintillation fluids in our country, it turns out that the amount of radioactive carbon was one curie compared to 230 million curies in the atmosphere. If it were to be burned, I can not believe that one in 230 million curies is going to have a terrible biologic effect. Now the amount of tritium that we use in New York (even though Leonard Solon will not permit us to burn it) represents 1% of the tritium in rainfall. I can not believe if it were 2%, and I were off by 100%, that anything terribly serious would happen.

UNIDENTIFIED SPEAKER: What if the people said that they didn't like it. Do you think that you should make the. . . .

YALOW: I am living in a democratic country, and this is the point I am making. Science is independent of policy. And where scientists get confused is they tend to say: "Well, we will make our extrapolations in an attempt to influence policy." That is not what I am doing. I am saying that we have an excess of scientific data demonstrating that the effects are so low that we simply cannot measure them. So that even if I am wrong and the effect is not zero, it is still a nonmeasurable effect. Now after I say this, in a democratic country, people can vote anything they feel like for any particular purpose. There is no way a scientist should say that policy decisions must be based solely on science. However, the laws of science do not change because people do not like their implications.

UNIDENTIFIED SPEAKER: Isn't a latent cancer nonobservable? By your definition?

YALOW: A latent cancer is not nonobservable because when observations continue for a very long period of time we would have observed effects from increases in natural background radiation.

UNIDENTIFIED SPEAKER: You talked about the nonobservable effects—a person with latent cancer has no observable effects. Does that make it a non-cancer?

YALOW: Yes, if it is latent and does not hurt him; the answer is yes from a practical point of view. I ended up saying that 27% of the Japanese have carcinoma in situ in their thyroids. They have what you would call a latent cancer in their thyroid.

UNIDENTIFIED SPEAKER: About nuclear energy, I would like to say that in the United States today, nuclear energy only furnishes 13% of our energy, only 3% of our electricity. The Harvard report has shown that with proper conservation, we need not import any oil from the Middle East. The radioactive waste of nuclear energy plants throughout the world can produce 25,000 nuclear bombs per year. When you think you have 50,000 bombs and you think of the possibility of nuclear war which can be utilizing the nuclear wastes we have, we are talking about something really significant. I don't think our problem is minimal radioactivity.

YALOW: I just want to say that our thesis tonight is low-level radiation effects, and legitimately this is what we are supposed to deal with. Now if you want to have a discussion on nuclear power, then I think you should have it some other time. And if you want to say that low-level radiation effects are the same as nuclear war, I think this is a distortion of the truth. It may be exciting to some of you to think that low-level effects mean nuclear war. It is not exactly the same thing.

UNIDENTIFIED SPEAKER: I do not want to take sides on the question of whether low-level radiation is a health hazard or not. I think that we are likely to see, according to what has been discussed in this room tonight, a significant increase in nuclear activities in the next 20 or 30 years. And the question is what are we going to do about it? Should we worry about it at all?

YALOW: I would worry about it. Now would you like to suggest that we have unilateral disarmament?

UNIDENTIFIED SPEAKER: I am not talking about disarmament. I am asking what policies should be formulated to deal with the increasing radiation that we all think we will be faced with? What the hazard will be, if any?

YALOW: I would say that the only real hazard from radiation is nuclear war.

UNIDENTIFIED SPEAKER: In other words, you think that as we increase nuclear energy usage, radioactive waste disposal and so forth, there is no hazard in the foreseeable future? Is that what you are saying?

YALOW: People say that scientists are equally divided on this. In fact, scientists are not equally divided on this. The Physical Society, the Health Physics Society, the engineers support nuclear power development. Few of them are working in nuclear energy. It isn't that they are all making their money out of it as Mancuso makes his money out of government funding for studies of effects of low-level radiation. It is the general considered opinion of the scientific community expressed not only through the National Academy of Sciences report on Energy in Transition, but in numerous publications, that the way to go for the next 25 years is coal or nuclear power. These are in effect the only options that we have open to us. We know something about the dangers of nuclear power and this has been calculated and recalculated. We know in part the dangers of coal. Not completely, because we really can not calculate the acid rain and greenhouse effect and that sort of thing. It is abundantly clear to everybody who really knows that solar energy simply does not have the density to deliver the amount of energy we need for the increasing population in this world. The ultimate problem that we all face is the increasing number of people in the world and their increasing aspiration to have what we have. And the only way in which we can really generate enough food is through power. We need power for fertilizers, we need power to bring water in, we need power in order to produce enough foods for us to live.

Public Health: Priorities and Policy Setting in the Real World[a]

ANITA S. CURRAN, AVIS W. EFFINGER,
ERNESTINE S. PANTEL

Westchester County Department of Health
White Plains, New York 10601

and

JOHN P. CURRAN

Department of Pediatrics
Lutheran Medical Center
Brooklyn, New York 11220

Public Health has been defined in many ways. When these definitions are arranged chronologically, they present a word picture of the evolution and development of the field. In the Sixth Edition of *Public Health, Administration and Practice,* Hanlon writes:

> Early definitions limited public health essentially to sanitary measures invoked against nuisances and health hazards with which the individual was powerless to cope, and which, when present in one individual could adversely affect others. Insanitation and later communicability were the criteria followed in deciding whether a problem fell within the concern and jurisdiction of public health. With the great bacteriologic and immunologic discoveries of the late nineteenth and early twentieth centuries, and the subsequent development of techniques for their application, the concept of prevention of disease in the individual was added. Public health then came to be regarded as an integration of sanitary science and medical science.[1]

And in 1847 Solomon Neumann recognized that "Medical science is intrinsically and essentially a social science, and as long as this is not recognized in practice, we shall not be able to enjoy its benefits, and shall have to be satisfied with an empty shell and a sham."[1] But it is only recently that medicine and public health have become widely recognized as applied social sciences.

A comprehensive, and probably the most widely accepted, definition is Winslow's, in an article published in 1920: "Public Health is the Science and Art of (1) preventing disease, (2) prolonging life, and (3) promoting health and efficiency through organized community effort for

(a) the santitation of the environment;
(b) the control of communicable infections;
(c) the education of the individual in personal hygiene;
(d) the organization of medical and nursing services for the early diagnosis and preventive treatment of disease, and,

[a]This paper was presented at the October 1, 1980 meeting of the Section of Science Policy of the New York Academy of Sciences.

0077-8923/83/0403-0061 $01.75/2 © 1973, NYAS

(e) the development of the social machinery to insure everyone a standard of living adequate for the maintenance of health.

so organizing these benefits as to enable every citizen to realize his birthright of health and longevity."[1] Early definitions reflected simpler times, environmental and political circumstances with which the public health professional was confronted. In the 18th and 19th centuries we functioned in a much less complex milieu, with fewer pressures of multi-factorial nature. There was less scientific knowledge, less technology, and the public had lower expectations.

Since the beginning of this century we have experienced the most remarkable scientific, technologic, ecologic, sociologic and economic upheaval in history. Advances by virologists, geneticists, physicists, chemists, toxicologist, immunologists and engineers have thrust upon us a bewildering number of new weapons, systems, and methods of prevention and treatment. And, at the same time, these advances pose new questions and often create new problems, uncertainties, and fears. We have become increasingly aware of the cost of this progress—abandonment of clean water, clean air and open spaces, exposure to toxic wastes and other man-made environmental hazards; encroachment of human habitation on public watersheds; urban blight, overpopulation, noise pollution, and all the other daily stresses that strain our sanity, and tip our emotional balance.

Along with this extraordinary explosion in scientific knowledge, a large number of power bases, each with its own agenda, have emerged.

In the 1980s, a fundamental question is whether public health policy will be made on the basis of scientifically derived knowledge, *or* in response to the many pressures to which the decision maker is subjected, *or*, as I believe, on the integration of scientific knowledge with political and economic realism and a "good seat of the pants" understanding of health systems management. In other words, public health policy, while based on scientific fact, must be established and implemented within the framework of the political structure. And because there are inherent differences between the scientific and political processes, this often leads to conflict, misunderstanding, and, occasionally, open warfare between the scientific and political communities.

These differences between science and politics as disciplines was nicely described by James Merck, Professor of Political Science at Stanford University, in a speech before the National Academy of Sciences, when he stated:

> Science presumes a process by which alternative theories are evaluated systematically against available data . . . in order to rank ideas in terms of their *plausibility*. Politics presumes a process by which alternative theories are compared on the basis of the power of people supporting them, in order to rank programs in terms of their *acceptability*. One process attempts to *reduce* subjectivity through standardized procedures designed to assure verifiable knowledge; the other attempts to *organize* subjectivity through a set of bargains designed to assure social stability. One seeks *data;* the other seeks *allies*.[2] (emphasis mine).

The public health physician who is health officer for a major political subdivision (city, county, or state) has the responsibility for setting priorities, and establishing, interpreting, and implementing public policy. And as members of the scientific community, health officers seek and need data; but, to be effective, they also seek and need *allies*. Under the county charter, a Commissioner of Health is designated as the executive officer responsible for carrying out the mandates of the New York State Public Health Law and the State Sanitary code. Chapter 261 of the Administrative Code of the Westchester County Charter stipulates that "the Department shall have jurisdiction to regulate all matters affecting public health in the County Health District."[3] The State Public Health Law defines broad responsibilities for local health officers.[4] Responsibilities and powers are further defined through 1000 pages of the State Sanitary Code,[5] which includes specifications on dissemination of information, such as the causes, nature and prevention of prevalent diseases; the preservation and improvement of health; the collection of information on reportable diseases; registration of births and deaths; enforcement of all provisions of the Public Health Law and the Sanitary Code; and the employment of staff necessary to enable the department to carry into effect those responsibilities, and the many orders and regulations of the county Board of Health.

All of these duties and responsibilities are discharged in the pressure cooker of public life, which is constantly fueled by a variety of often contentious power bases . . . the political structure, news media, regulatory agencies, the scientific community, industry, special interest groups, and the public at large. An now, more than ever before, all actions and decisions, with a few specific exceptions, must be made in full view of any interested parties as a result of the Freedom of Information Act.

In the political arena, a politician's behavior is often influenced by the need to gain and retain elective public office. County legislators, like members of the House of Representatives, are elected to terms of only two years, which means that they are always running for re-election. And in our political system a politician presents himself to his electorate as a pro or con on a particular issue (e.g. abortion), or as a yea on nay on legislative matters (e.g. parents' rights to be informed with regard to abortions of minor children). In recent years some politicians have achieved public office by exploiting a single, volatile, emotional issue, including the public's fears, misconceptions and unwarranted expectations in the area of public health. Also, David Mechanic believes some politicians tend to avoid confronting antagonistic groups, and political interventions often consist not of bold and innovative thrusts that make significant impact, but rather a sort of tinkering with the fringes of already existing policy.[6]

For the news media, the pressure to sell papers and attract listeners and viewers are immediate and intense. This is compounded by the reporter's need for recognition and career advancement, which may lead to the blowing out of proportion of small issues, lending them importance beyond their worth, with headlines designed more to inflame than inform. The pressure to break the story first can lead to incomplete investigation, and publication without verification.

Federal and State regulatory agencies play an increasingly large role in the health field. These agencies attempt to regulate the quality and quantity of care delivered, where and under what conditions these services will be provided, and at what cost.

In the area of environmental health, regulations seek to protect the public from the ever increasing pollution of the environment that results in degradation of the air, water, and land. Occupational health has become a specialty with its own regulatory agency geared to protecting the health and safety of the worker in the work place. Even though environmental and occupational regulations hold great promise as the source of health improvement and consequent reduction of medical care costs, may politicians and industrialists are wary of the large compliance costs that such regulation impose. All branches of government are reviewing the benefits of health standards vs. their costs. This approach has merit. However, in the quest for simple, quantitative measures of costs and benefits, a cost-benefit analysis is likely to underestimate the health benefits of regulations, and overestimate their costs.

The scientific community provides the knowledge that forms the foundation upon which all public health programs are constructed. However, many scientists are not content with simply presenting the facts, but wish to become more involved in the decision-making process. This desire may stem from personal commitment, or from the desire for personal or professional recognition, or the need to survive in the highly competitive academic world. This involvement can have negative as well as beneficial implications for the formulation of public health policy.

Industry, with its focus on economic gain, has the resources to purchase the lobbying and scientific clout to refute threatening scientific data. For example, a new methodology for assessing health risks developed by EPA for the ozone standard is about 1/2 inch thick; the critique of that document prepared by an industry lobby group is approximately 2 inches thick.

The mobilization of special interest groups (labor unions, consumer activists, right-to-lifers, family-planning enthusiasts, and advocates for alternative energy sources, etc.) has created competing constituencies to which the public health officer must respond. Each special insterest group defines a "problem," real or perceived, around which it rallies. These groups are characterized by their beliefs in the seriousness of the consequences of some defined event, and the likelihood of its occurrence.[7] They have learned to take advantage of the complexities of the decision-making process by "using the special expertise and the political legitimacy that they have attained to exercise greater dominance over the system than their numbers, special wisdom, or degree of neutrality would justify."[6] The issue of personal liberty is often raised, and by making a pitch for individual freedom, special interest lobbies can influence legislation and regulations in ways that are not only detrimental to public health, but that ignore the concept that one man's freedom is often another man's burden (increased taxes and insurance premiums). An example is the recent history of motorcycle helmet laws in this country. In 1976, Robertson shows that enactment of motorcycle helmet laws was associated with a 30% reduction in deaths among motorcyclists.[8] Ignoring this information, 27 state legislatures subsequently revoked these same laws, producing an increase in motorcyclist mortality. It was as if scientists, having found a successful treatment for a disease, were impelled to further prove its efficacy by stopping the treatment, and allowing the disease to recur. Recent studies show that when helmet laws are repealed, helmet use drops by about 50% and motorcyclist deaths increase by 38 percent. In the face of substantial evidence of the effectiveness of this law, and a poll showing strong voter support, as well as the negative effects of its repeal, a Maryland Senate Committee voted overwhelmingly not to reinstate the helmet law. This demonstrates that the combination of solid data, plus the pressure exerted by public health, medical, safety, and youth groups, as well as state agencies (transportation, police, medical examiners), was not enough to counterbalance the enormous lobbying effort, primarily from ABATE (A Brotherhood Against Totalitarian Enactment) and the American Motorcycle Association. While extremely vocal, visible, and effective in their pleas to "let those who ride, decide," they do not represent the viewpoint of the majority of motorcyclists, much less the viewpoint of the general public. Yet, because of their successful lobbying, this year in the United States more than 1000 motorcyclists will die, and many more will be injured. The general public will share the burden of these deaths and injuries in a variety of impersonal ways; paying for acute and long-term care and rehabilitation, and through increased demands on limited resources, such as blood supplies, emergency services, and excess automobile liability insurance.

The individual freedom argument is neither new nor limited to helmet laws. It was used for decades to delay community pasteurization of milk supplies, and it continues to obstruct effective gun control legislation.

The level of sophistication of the general public regarding health care issues has increased. Yet, perceptions remain that conflict with established scientific facts and common sense. These has been a revolution of rising expectation; first, to demand a right to health care, and now to demand a right to health. The media, government and the health professions have fostered the concept that the public should be free of all disease and environmental threats. These expectations are real, but they are not realistic, or attainable. They have also led to the development of an extremely litigious society, increasingly suspicious of the public sector for a variety of reasons.

In many cases, it appears that the system is working against itself. The public health professional must contend with continued use of outmoded public health practices, such as efforts to prevent necessary and harmless childhood disorders such as chickenpox; the medical profession's persistent failure to subject the questionable procedure of tonsilectomy and adenoidectomy to a controlled study of its efficacy; the legalization of untested drugs, such as laetrile, as the result of pressures that have their major source in emotionalism and avarice; the maintenance of health care facilities that are substandard and unnecessary simply to provide employment to a local constitutency or satisfy other than health care needs; health insurance programs that will pay for circumcisions, but not removal of an extra finger, since the latter is considered to be cosmetic surgery; a system that condemns cigarette smoking as hazardous to your health and then provides government subsidies to the tobacco grower; and many others.

All public health activities must be accomplished in an atmosphere of growing fiscal constraints. The public is demanding, and getting, cost containment. The pressure of Proposition 13 has resulted in property tax reduction in many counties, including Westchester. The reality of enormous increases in health care costs as well as inflation further constrain the financing of preventive health care programs.

Americans spent $212.2 billion for health care in 1979, which was 9% of the gross national product.[9] This represents a 52% increase over the 1976 expenditure of $139 billion. Between 1950 and 1979, total health expenditures rose at an annual rate of almost 10 percent. Half of this increase was due to price increases alone.[10] For many years, about 25% of the health expenditures came from public funds. With the implementation of medicare and medicaid the proportion rose, reaching 43% in fiscal year 1979.

With the rapid increase in expenditures for medical care and the increasing proportion paid out of public funds, resource allocation decisions are critical. Yet, in this country today the private sector, with rare exception, and a large segment of the public sector are totally committed emotionally, philosophically and economically to the treatment of the sick, leaving only a small percentage of public resources for disease prevention and wellness promotion. Close to 96% of every federal health care dollar is spent on the care of the sick.[11] But there is much more to health than just not being sick. As the public agencies charged with preserving the health, in its broadest meaning, of the citizens of this country, health departments have accepted the responsibility for developing and implementing programs aimed at promoting health, preventing disease and prolonging meaningful life. As a consequence, health departments must take a good hard look at some traditional programs and do serious cost/effective priority setting.

I would like to give you an example of how these many factors interact and bring pressures to bear on the role of the decision-maker in the setting of priorities, and in the interpreting and implementing of public health policy.

In 1943, the Canadian Radium and Uranium (Can-Rad) Corporation began operations in the Mount Kisco area and supplied raw material for this country's first atomic bomb. It was part of the famous Manhattan Project. In 1957, the plant drew national attention because it was the subject of the first criminal prosecution for exposing employees to excessive radiation. After 1957, operations were scaled down, but continued until shortly before the building was demolished in 1966 to make way for urban renewal. The rubble from the building was removed and disposed of under permit from the State Department of Health and the area decontaminated in compliance with Department of Labor standards, and advice from the U.S. Public Health Service.

In March of 1979, a newspaper reporter researching the story of the Can-Rad Corporation obtained a geiger counter from a friend in private industry and surveyed the ground in the vicinity of the demolished factory. In a fenced-in area used for dead storage behind an abandoned freight depot, she discovered a small area several feet square where the radiation levels were

0.35 to 0.4 millirems per hour. This has become known as the "Mt. Kisco Hot Spot." In actuality it would be more accurate to call it the "lukewarm" or "tepid" spot, since it violates no emission standards and does not qualify as a health hazard. Yet the combination of the interests of the reporter, the timing of the discovery right after the incident at Three Mile Island, the public's fear of the dangers of radiation, the local politicians desire to be responsive to perceived threats to the community, the failure of the State to cooperate as promised, and my failure to respond immediately and decisively to what I considered, at the time, to be a minor issue, has resulted in the blowing up of a non-health problem into a media event. As a consequence of the converging pressures the Village, the County, and the present landowner decided to combine their resources to remove the contaminated soil. Two separate contracts have been awarded for the decontamination of the land. One agreement between the County of Westchester and the contractor calls for the removal of approximately 5 cubic yards of radium 226–contaminated soil with the County paying for health physics technicians and general laborers to perform these services. In addition, the County will pay for soil analysis to determine radioactivity levels for packaging and labeling purposes. The second agreement is between the Village of Mount Kisco and the same contractor for approval disposal containers, transportation, and burial at an approved site.

You may be asking at this point who would want this job? Hanlon writes that

> it is significant that many outstanding persons who otherwise might have gained greater material benefits have devoted their lives to the public health movement. One of their characteristics has been a special enthusiasm for their work. What makes such ardent proponents of those who have given sincere thought and effort to this field? Does it present some unique challenge? Perhaps here more than elsewhere man at last has the opportunity successfully to adapt the creatures and environments of nature to himself and his welfare, rather than to submit helplessly to them. But, one might ask, what can a sincere and honest person hope to accomplish in the face of superstition and ignorance of the large part of the people, political interference, inadequate funds, and sometimes personnel of poor quality? Is not impatience justified?[1]

And the Irish parliamentarian, Edmund Burke, in the face of repeated criticism, observed: "Those who carry on great public schemes must be proof against the most fatiguing delays, the most mortifying disappointments, the most shocking insults, and what is worst of all, the presumptious judgments of the ignorant."[1]

Who would want this job? I would!

Because this is an exciting, stimulating, dynamic, creative job. All of the conflicting forces and power bases—the media, regulatory agencies, politicians, the savvy public, the scientific community, industry, and special interest groups—can all be of assistance to the public health professional if he or she is willing to get out there and meet these forces head on, and ust them to benefit the public.

The politicians can be of immeasurable assistance in allocating the funds and securing the resources necessary to implement public health programs.

The news media are often the only source of health education information for certain segments of our population. The media can be an inestimatable ally in publicizing our message.

Federal and state regulatory agencies can supply the necessary "clout" to finally get the job done. Witness the new State Education Law requiring "Shots for School"—virtually 100% immunization against certain communicable childhood diseases is about to become a reality.

The scientific community can offer the backing needed to give validity to the data on which regulations are based, refute false accusations, influence the public to support our programs, and provide the moral support needed to "carry on."

Industry can be a tremendous support in providing "seed money" to establish innovative programs, and they do.

Special interest groups and the knowledgeable public are often motivated to support public health activities, and can thus plead our case in the political arena.

This is where the action is—you are on the firing line, you are the point man, so to speak, where all those conflicting forces converge. The public health practitioner must extract all the benefits from the above, while resisting the sometimes dehumanizing influences of the scientific explosion, realizing that public health is, and must remain, a deeply social as well as a scientific discipline. And because of the complex nature of its interests and responsibilities, a highly politicized one.

REFERENCES

1. HANLON, J. J. 1974. Public Health Administration and Practice. 6th edit. C. V. Mosby. St. Louis, MO.
2. MARCH, J. 1979. In a speech before the National Academy of Sciences, 1979, as quoted in D. AXELROD, Health Promotion and Disease Prevention—Priorities in the 80's. Speech at the 30th Annual Meeting of the New York State Public Health Association. Buffalo, NY, June 10, 1980 (Italics aded.)
3. Westchester County Charter and Westchester County Administrative Code. 1979. General Code Publishers Corp. Spencerport, NY
4. McKinney's Consolidated Laws of New York (annotated). 1979. Book 44, Publishing Co. Saint Paul, MN.
5. STATE OF NEW YORK. Official Compilation of Codes, Rules and Regulations. 1969. Department of State, Albany, NY.
6. MECHANIC, D. 1979. Future Issues in Health Care. Social Policy and the Rationing of Medical Services. The Free Press. New York.
7. MECHANIC, D. 1974. Politics, Medicine and Social Science. John Wiley & Sons. New York.
8. BAKER, S. P. 1980. On lobbies, liberty and the public good. Amo J. Public Health **70**(6): 573–574.
9. BRAZDA, J. F., Ed. 1980. Washington Report on Medicine and Health. McGraw-Hill. Washington, D.C. **34**(34): 4.
10. UNITED STATES DEPARTMENT OF HEALTH, EDUCATION AND WELFARE. Health, United States, 1976–1977. DHEW Publication no. (HRA) 77-1232.
11. MCGINIS, J. M. 1979. In a speech at the National Association of Counties Annual Meeting, Kansas City, MO. July 18, 1979.

Science and The "Regulation" Bogey[a]

JOHN E. ULLMANN

Department of Quantitative Methods
School of Business
Hofstra University
Hempstead, New York 11550

It is most appropriate to start with the dictionary definition of "bogey" as a "specter" or "phantom" or, as in "bogeyman," a "monstrous imaginary figure used in threatening children." The regulations under discussion are those established by the federal, state and local governments to implement a variety of laws. Let that be noted carefully: They are a tool of law enforcement and they exist because the laws exist. And the laws exist because they were passed by the legislative process and withstood challenges, if any, in the courts. It is necessary to state this at the outset, for the impression is sometimes created that regulations spring full-blown from the head of some "bureaucrat." Such views were certainly part of the use of opposition to "regulation" (usually left undefined) in bringing a conservative political message to the American electorate in the Presidential election of 1980.

Well, it worked. A set of issues of increasing complexity was reduced to simplistic sloganeering. Nowhere in the process was there any discussion of the needs for, and structure of the kinds of regulation that were the subject of the attacks; least considered of all were the essential historical factors in the controversy. The basic confusion of the electorate is shown by the fact that all this took place, even though public opinion polls showed strong support for consumer and environmental protection, for safety in the work place and for other objectives of the self-same regulations. As the Reagan cabinet appointments made their way through the rumor mill of the transition and were ultimately announced, one could almost hear the loudest complainers in the business community "smacking their lips in a refined manner, like a fox in a henhouse," to borrow a phrase from S. J. Perelman. For many scientists, the specific political and scientific struggles portended by the success of this tactic imply a whole range of professional opportunities, as well as threats and dangers. An examination of the issues is therefore clearly urgent.

It is unfortunately tautological to say that the United States, like most of the rest of the world, is facing difficult times ahead. In any society confronted by such a prospect, the central issue is who will pay for these troubles, since burdens of this sort are almost never allocated equitably. In this atmosphere, regulations, particularly of business, have become the object of ever more strident political conflict. The chief complaints are that they stifle economic development and thereby cost jobs, that they raise prices or seek to foster

[a]This paper was presented at the December 10, 1980 meeting of the Section of Science and Public Policy of the New York Academy of Sciences.

0077–8923/83/0403–0069 $01.75/2 © 1983, NYAS

social objectives with which the complainer disagrees. They are also alleged to be burdensome and cause inordinate delays.

Nobody would deny that there is substance to the latter complaint. Government agencies, like most large organizations whether public or private, spend much of their resources on internal memo-sending, rather than on discernible service to the outside customers. They appear to follow a version of the First Law of Thermodynamics, in which resource input brings an ever smaller proportion of useful work as compared to "internal energy," i.e. the churning about referred to above. But it is an entirely different matter to argue that the subject of the regulations should therefore no longer be a public or governmental concern.

The history of regulation in the United States happens to be one of reluctance of government to become involved in the regulatory process. It has done so only under extreme public pressure following manifest failure of voluntary remedies or so-called self-regulation. When it finally moved, much of the enabling legislation was made so complex that, as will be seen later, it spawned an "industry" of sorts to deal with the regulatory processs.

To scientists in general and engineers in particular, the role of the business community is of central importance because science enters the economy and society to a large degree as a result of business activities or, at times, by a quasi-business carried on by a public authority.

Accordingly, one can group the concerns of scientists in three major categories. The first is the professional role of scientists in formulating policies and regulations regarding health and safety, with respect to products and processes and honesty in the marketplace. The second category involves the prospects of freedom of inquiry, not only for scientists but for all involved in the analysis of society and in creative work. Finally, there is the question of working conditions of scientists which in turn depend very much upon the quality of working conditions for everybody.

The above makes clear that not all of the large body of regulations as defined at the outset, is of concern to scientists as scientists. For example, the frequently criticized controls exercised by the Securities and Exchange Commission are mainly of interest to investors, even though some, as individuals, are scientists.

What an examination of the issues does show, however, is that one deals with highly complex scientific and social problems which render such questions as "Are you for or against regulations?" both simplistic and absurd. In fact, it is not difficult to show that the science-based regulations attempt to fulfill essential objectives and that, insofar as they have side effects, the cure is seldom worse than the disease; moreover, to carry the medical metaphor further, the side effects are often the result of the patient's poor attitude and downright misbehavior.

SCIENCE AS THE SOURCE OF REGULATIONS

In the above three categories the role of the scientist in matters of health and safety presents the widest and most heterogeneous set of issues. It also deals with problems of great antiquity, and it is instructive to see that modern regulations often only seek to do what has engaged human attention for centuries or even millenia.

For example, a recent and predictably shallow and self-serving article was done for the house organ of a major oil company[1] by the Center for the Study of American Business, which was then under the direction of Murray Weidenbaum, Mr. Reagan's designate as Chairman of the Council of Economic Advisers. Various cartoons illustrate it, among them a montage of rubber stamps, "approved," "violation," "inspected," and others. Among them is "Department of Weights and Measures." Now, as it happens, weights and measures have been regulated ever since they came into existence, i.e. since the Sumerian Empire. The Code of Hammurabi provided drastic penalties for violators and, indeed, gruesome punishments for them were meted out in many law codes since then.

In medieval Vienna, a special ducking stool was used for bakers who short-weighted their bread; they were dunked in the Danube. One wonders how the enforcers of that regulation would feel about the so-called "hungry pound" of fifteen ounces which is now used for loaves of bread, especially in the Western United States. In a similar example, the liquor industry was faced with the need to use metric sizes in their bottles and settled on 1.75 liters as a replacement for the old half-gallon bottle when 2 liters, as used for soft drinks, would have been the obvious choice (the old quart liquor bottle became a liter). However, the 1.75 liters is 59.2 fluid ounces rather than 64 in a half gallon. In both bread and liquor, the cut in size without, as invariably happened, a cut in price, meant a silent price increase of about 7 percent. So what is being asked here? A legalized "butcher's thumb" on the scale?

The continuity of issues is demonstrated with equal clarity throughout the areas of health, safety, and the environment. Within an American context, it is appropriate to start at the beginning of the twentieth century when regulatory activity in the modern sense first began. It was not only the time of the first food and drug regulations, but also of the anti-trust laws, legally mandated cuts in working hours in a few places, and of utility regulation, which had its origins in the establishment of the Interstate Commerce Commission in 1892. Before the first drug laws, heroin was used in cough medicine, and opium and lots of alcohol were staple ingredients of the ubiquitous patent medicines of the time. The drug sections of the Sears Roebuck catalogs of the time, which were reprinted a few years ago, make alarming reading.

A novel of the period (1905), Upton Sinclair's *The Jungle,* showed that fiction containing factual (and easily provable) reportage can become a powerful stimulant of sociopolitical action. Sinclair sought to expose both the filth served up to the public by the meatpacking industry of the time, and the dreadful condition of its workers. In the event, only the inspection laws were tightened, which caused Sinclair to remark some time later: "I aimed for the public's heart and I hit it in the stomach."

Still, Sinclair's accounts establish something of a benchmark and are thus relevant to the current discussion in that they shed a rather harsh light on the preregulatory "good old days." For instance, Sinclair writes: "The children [of his protagonist] were not as well as they had been at home [in Europe] . . . How could they know that the pale blue milk that they bought around the corner was watered, and doctored with formaldehyde besides?"[2]

As it happens, formaldehyde, while no longer found in milk, is a highly controversial substance even in 1980. Where urea formaldehyde insulation is installed by using the foamed-in-place method, formaldehyde fumes have been known to cause headaches, nausea, respiratory and visual disturbances and persistent headaches, often lasting for the several months required for the fumes to dispel. There is also some evidence from animal studies that formaldehyde may be carcinogenic. Naturally, the industry counters all of this by focusing on denials of carcinogenicity; however, the other symptoms surely are disagreeable enough to call the further use of this material into serious question.

Sinclair continues:

> How could they find out that their tea and coffee, their sugar and flour, had been doctored; that their canned peas had been colored with copper salts, and their fruit jams with aniline dyes? And even if they had known it, what good would it have done them, since there was no place within miles of them where any other sort was to be had?

Again, the issue can hardly be considered settled; food colorings remain matters of controversy. Red dye No. 3, for instance, has been banned and such other products as the coloring agents in hair dyes are believed to have at least some carcinogenic effects. Industry reaction to such allegations is to change the offending molecule very slightly, requiring the whole complex and long drawn-out process of cancer testing to start all over again.[3] Also, let it be noted, powerful industry groups still object to labeling of food ingredients, including in their advocacy such condescending (and sexist) assertions that the labels take too long for the housewife to read.

Sinclair writes of smoked sausage, a staple of his protagonist's diet, as being adulterated with "potato flour," a cellulosic residue of the potato after the starch and alcohol have been extracted.[4] Having been banned in Europe it was exported to America. Sausages still retain their often unwelcome and controversial mysteries. Sodium nitrates and nitrites are considered risky as preservatives, though alternatives also present problems. Here and in Europe, the use of mechanical deboning is attacked by consumer groups, because it laces the sausage meat with non-nutritious cartilage and other materials that only add to the bulk. Should the products contain sugar or dextrose, dried milk or soy solids? How much fat should there be?

But adulteration, now as then, also extends to foods that, one would suppose, are by nature more or less pristine. Recently, for instance, the use of "cold water starch" in tomato sauces and other fruit and vegetable products was hailed by the trade press of the food industry as a step forward in product development[5]; but perhaps one should not be surprised at such value judgment from an industry that has, by selective breeding, managed to give tomatoes the consistency, and sometimes the taste, of tennis balls.

As to dumping unsatisfactory products abroad, the practice is likewise evidently alive and well. In one case, in 1980, an English woman traveling in

Mozambique took an over-the-counter analgesic drug which was based on amidopyrine. This, like its close chemical relative antipyrine, is no longer permitted in the United States and in most of Western Europe because one of its side effects is the irreversible and sometimes fatal blood disease agranulocytosis; the woman was stricken and suffered permanent health impairment. Yet, at least five companies, including major multinationals such as Ciba-Geigy, Sandoz, and Hoechst, as well as the Polish government pharmaceutical monopoly sold this compound in Mozambique.[6]

Such incidents will, however, do little to diminish the steady drumfire of the U.S. pharmaceutical industry against the requirements for drug safety and efficacy that allegedly "prevent" it from giving us great new medicines. Still, it was the much maligned U.S. Food and Drug Administration that delayed the approval of thalidomide long enough to make its horrendous side effects clear.[7] Given the national habit of pill-popping, it is hard to underestimate the scope of the disaster that was thereby avoided.

Working conditions likewise present extensive scientific and technical issues. Here is Sinclair on the meat packers who are the subject of his book:

> Some worked at the stamping machines, and it was very seldom that one could work long there at the pace that was set and not give out and forget himself and have a part of his hand chopped off. There were the "hoisters" as they were called, whose task it was to press the lever which lifted the dead cattle off the floor. They ran along upon a rafter, peering down through the damp and the steam; and old Durham's architects had not built the killing room for the convenience of the hoisters, and every few feet they would have to stoop under a beam, say four feet above the one they ran on; which got them into the habit of stooping, so that in a few years they would be walking like chimpanzees. Worst of any, however, were the fertilizer men, and those who served in the cooking rooms. These people could not be shown to the visitor, for the odor of a fertilizer man would scare any ordinary visitor at a hundred yards, and as for the other men who worked in tank rooms full of steam and in some of which there were open vats near the level of the floor, their peculiar trouble was that they fell into the vats; and when they were fished out, there was never enough of them left to be worth exhibiting—sometimes they would be overlooked for days till all but the bones of them had gone out to the world as Durham's Pure Leaf Lard![8]

While such physical necessities as adequate guard rails are now more carefully designed than in Sinclair's day, the prospects for health damage in the course of one's work remain appallingly plentiful. Exposure to deleterious chemical agents is an ever increasing hazard, although here too, the precedents go back over many decades. One hardly needs to be reminded of the effects of asbestos, of hydrocarbons, of heat, speedups, dusts, noise and fumes. They are extensively covered in the literature.[9,10]

While the social costs of industrial accidents and sickness continue to pile up at increasing rates, few parts of the anti-regulation campaign have caused as much heat as those against OSHA, the enforcing agency for the Occupational Safety and Health Act. There is, in fact, an organization called "Stop OSHA," established under the auspices of the U.S. Chamber of Commerce. While a candidate, President Reagan proposed the abolition of OSHA, although he pulled back somewhat from that position later on.

When the agency was first created, it did indeed misinterpret some long accepted engineering handbook data and also based some of its rules on obsolete standards of the American National Standards Institute (ANSI, formerly ASA),[11] but these early problems have been largely corrected. ANSI standards, incidentally, constitute an enormous body of "regulations," if one is willing to call them that. They are created by an organization supported by some 900 private firms, trade associations, and public agencies. They have to be based on consensus and, moreover, have in recent years required inputs from the consumer's side as well. ANSI standards are thus hardly "imposed" from the outside or by those famous "bureaucrats"; yet the more strident opponents of OSHA often ignore the distinction.

As to environmental damage, there are many places where one can change Tacitus' condemnation of some of the actions of his fellow Romans,[12] "They made a desert and called it peace" to "They made a desert and called it development." Again, the problems form a melancholy continuum in human history. The deforestation of China, ancient Greece, North Africa and other areas have left a legacy of decline and at times starvation, which has lasted until our time and is likely to continue for the foreseeable future. At present, the American sunbelt subsists in large part (about 40% in the case of Arizona) on its capital of ground water.[13] One day, at present rates, the pumps dipping into the aquifers will run dry and then the area will become one of Tacitus' deserts.

Meanwhile, on his appointment as Secretary of Agriculture on December 22, 1980, J.R. Block sternly declared that "food was America's greatest diplomatic weapon."[14] Yet, only shortly before that, his own Department had published most alarming reports on the acceleration of soil erosion and destruction of agricultural land, as the result of pushing from even higher food production while ignoring such elementary conservationist regulations as contour plowing.[15] This damage seems certain to be compounded by the prospective policies of J.G. Watt, the chosen Secretary of the Interior, who had made a career out of fighting conservationist regulations in national forests and lands.

Scientists are involved in all of these issues as advocates on both sides or all sides if there are more than two. As noted, every new regulatory area appears to breed an industry of sorts, consisting of those involved in the adversary proceedings, litigations, and analyses that go into them. One such group of establishments prepares environmental impact statements for industry. Some of them remind one of Samuel Johnson's remark that "in lapidary inscriptions, a man is not upon oath"; they are, in short, advocacy, rather than fact finding and, of course, there is yet another group that has to try to refute them.

Business waxes wroth over these impositions, but they do generate a great many jobs. In their classic examination of bureaucracy, Blau and Meyer point to a symbiosis of sorts between governmental and private bureaucracies:

> Capitalism . . . promotes effective and extensive operations of the government. It also leads to bureaucratization in other spheres. The expansion of business firms and the consequent removal of most employees from activities directly governed

> by the profit principle make it increasingly necessary to introduce bureaucratic methods of administration for the sake of efficiency . . . Strange as it may seem, the free enterprise system fosters the development of bureaucracy in government, in private companies, and in unions.[16]

One reason for the existence of bureaucracies is in fact the uniform adjudication of a variety of problems. Historically, bureaucracy has always supplanted ad hoc attempts, usually by feudal lords or other local chieftains, to legislate, which they often did capriciously and inconsistently within a country.

Once they are established, the cutting back of bureaucracies has always proved extremely difficult. Such steps as the deregulation of the railroads, trucking, and airlines has brought a surprising degree of anguished reaction from those previously engaged on *all sides* in the administration of regulations now no longer applicable.

One essential part of the elaborate games of tournament and tilt that have grown up around the regulatory process is for one or other contending party to follow the adage, "when losing the game, change the rules." During the 1979–80 Congressional debates that ultimately led to legislation sharply restricting the powers of the Federal Trade Commission, specific clauses were attached exempting particular industries with sufficient political clout from the attentions of the Commission. One is again uncomfortably reminded of Tacitus'[17] comment on the last years of the Roman Republic: "And now laws were made not for the public only but for particular men, and in the most corrupt period of the commonwealth, the greatest numbers of laws were made."

Finally, much of the opposition to regulation is clearly a result of the inability of managements to meet the standards set even where these clearly satisfy a public need. In that sense, the opposition is a mask for managerial incompetence. It is easy to blame environmentalists, consumerists, and those concerned with worker safety for the inability of business to cope with its problems.[18] Yet, business, especially of a technical nature, must be essentially a problem-solving process if it is to continue. The decline of innovation, the failure to improve productivity, and the decline in quality have been amply discussed in the literature and are reflected in the recent concern within the United States with what is called "reindustrialization." The decline is actually the result of tilting the priorities in investment and the allocation of technical talent towards military and space activities and away from the maintenance of what might (perhaps in a rather old-fashioned way) be called the industrial patrimony.[19]

In spite of, or perhaps because of, the poor stewardship of a vital basis of society, the reaction of business goes far toward answering a question raised at the beginning of this paper, namely, "Who will pay for the difficult times ahead?" *Business Week*[20] puts it as follows:

> The leaders of the various economic and social groups that compose U.S. society should agree on the program for reindustrialization and present that program to Washington . . . Business, labor and academic leaders should establish a forum to hammer out a new social contract for the U.S. Special groups must recognize that their own unique goals cannot be satisfied if the U.S. cannot compete in

> world markets. The drawing of a social contract *must take precedence over the aspirations of the poor, the minorities and the environmentalists.* (Emphasis added.)

In a later editorial it continues:[21] "It also means turning off the continuing attacks on business that are the stock-in-trade of politicians and self-appointed consumerists." This, of course, sounds uncomfortably like a concerted attempt at repression of unpopular opinions. One might also raise the point of order that a corporation, too, is almost by definition "self-appointed." Any elementary textbook in management makes clear that a corporation defines its own objectives and purposes. More generally, one of the fortunate aspects of a free society is precisely the wide range of activities to which one may appoint oneself. Meanwhile, the wistful pleas for business to be "unleashed," which are also part of the anti-regulation campaign, appear to have as much reality behind them as the suggestions in the 1950s for "unleashing" Chiang Kai Shek's army against the mainland.

FREEDOM OF INQUIRY

Of all the misperceptions in the anti-regulation campaign of 1980, the crassest was undoubtedly the carefully fostered notion that an end to "regulation" would "get government off our back." Yet, aside from gun control which was stoutly resisted, the conservative coalition that gained power includes influential groups that would vastly *extend* government authority, not merely in the regulation of wide areas of personal conduct now left very properly to individual choice, but in the direction of censorship and increased restrictions on scientific enquiry. Where science impinges on the business activities described above, the impediments to independent analysis seem especially likely to grow.

Scientists face increased dangers in at least four respects. First, there is the conflict between science and religion. Mr. Reagan's espousal of "creationism" as a scientific theory deserving equal time with evolution, promises, if implemented, to revive the Scopes trial and the prior conflicts of the 19th century. H.L. Mencken's descriptions of the trial[22] make disturbingly topical reading. Martin Gardner[23] points out that "the Catholic Church has reacted to evolution with considerably less frenzy than the orthodox Protestant groups . . . Perhaps the Church had learned a bitter lesson from her experiences with Galileo." It is thus noteworthy that, as 1980 ended, the case of Galileo was being reexamined by the Vatican authorities amidst some quite substantial opposition from the more conservative clergy and laity. Today, it is discouraging to read, for example, Andrew D. White's classic account of these conflicts.[24] Reflecting the optimistic temper of his time and place—America in 1895—he really thought these controversies had been resolved once and for all in favor of scientific freedom.

A second danger is that legislatures, swayed in part by that kind of pseudoscientific thinking, will seek to repeal scientific fact, much after the fashion of Rudyard Kipling's "The Village that Voted the Earth Was Flat."

The legitimization of Laetrile is a case in point, as were political pressures applied on behalf of earlier nonconventional cancer treatments such as Krebiozen and the Hoxey Clinics.[25] This is not to be taken as an endorsement of "establishment" medicine; to the contrary, but treatments of whatever kind must meet criteria of proper testing and efficacy, of merchantability, to use the legal term. Such characteristics are expected of the humblest products one buys; why then should vital medical questions deserve any less and why should the state, in effect, put its weight behind some favored group's unproven advertising claims?

The third danger, sometimes stemming from the second, is that individual politicians seek to make capital out of assaults on specific scientific projects, for ideological or religious reasons. The latest in a long succession of Congressional figures who have attempted to do that is Senator William Proxmire (D-Wisconsin) with his "Golden Fleece of the Month" award. As a result of this he was sued by one of his victims, Professor Ronald Hutchison of Western Michigan University. Though Proxmire lost his case, the legal costs borne by his opponent turned out to be ruinous.

Most important of all perhaps is the restriction that is placed on science by what government and major corporations decide to support. Given the capital costs of even "basic" research—assuming it can be defined—who gets grants determines to a large extent the science and technology that society receives. Once again, the military and aerospace priorities in research bulk large. On a more general level, the transition team of the Reagan administration has proposed a more application-oriented role for the National Science Foundation. Clearly, if implemented, this will have far reaching effects on who can do what.

One other issue which has received attention in this connection has been the burdensome reporting provisions. It is true that in many cases researchers spend a disproportionate amount providing progress reports rather than dealing with the direct substance of their work. However, many of these provisions are put in precisely by those legislators who disapprove of the original objectives and therefore seek to burden the job with just that kind of work. When, later, the results are poor and slow in coming, the finger can be pointed at the purpose of the whole project.

Here too, things are not always what they seem. For instance the pharmaceutical industry and others often present huge stacks of documents required for approval of some product or project. Now, if these really are required, one can often make a case that less might be more. But then, what better way to show up the defects of "regulation" than by overloading the system with huge masses of paper?

WORKING CONDITIONS

Finally, scientists themselves must make a living. Obviously, there will only be jobs in areas that will be supported. However, even within such areas there arise important questions of professional integrity and ability to function in line with one's scientific findings and personal convictions. There is an extensive literature on this point and the Academy has hosted several conferences dealing with it.[26] It is beyond our scope to go into much detail, but there is every reason to be concerned over the prospect that only certain viewpoints on specific issues will receive support and attention. Referring to the operational difficulties discussed above, science is threatened less by those who set procedures than by those who seek to predetermine its conclusions.

Scientists must also be just as concerned as other workers with the laws and regulations that govern employment. The continuing viable functioning of OSHA is clearly in their interest, for scientists often deal with dangerous substances and processes. Equal employment opportunities are likewise of great concern, although there are several areas in science where the shortage of suitable talent is severe enough for prospective employers to be unable to follow traditional prejudices if they wish to fill their needs for good professionals. Such other laws as the protection of pension rights under ERISA also merit attention.

Finally, there are the other rules that govern security and conditions of employment. Here, it is particularly appropriate to note that the campaign against regulations has focused heavily on the job opportunities and working conditions of civil servants and other public employees. It has stressed such issues as their relatively greater security of employment, which contrasts sharply with the capricious manner in which people can lose their jobs in much of private industry. Unionization is likewise under attack. However, it behooves all who work to realize that employment standards are indivisible. If, as a result of the political pressures described here, the working conditions of public employees are significantly impaired, if public employees are used as generalized scapegoats for the ills of society, then working conditions for all will suffer.

TOWARD AN UNCERTAIN FUTURE

Scientists are deeply involved in the basis for policies on health and safety of products and processes and the protection of the environment. To the extent that their work will be increasingly subject to ideological constraints, there is the gravest danger that the scientific competence of the entire society will be reduced, perhaps fatally so. There are indications that this process is already under way. Science has been the great moving force behind the creation of modern life, with its horrors as well as its obvious benefits. The problems that will now have to be solved in order to maintain a society that can even remotely deserve to be called "humane" will require strenuous and costly efforts. In this context, the opposition to regulations in the areas discussed here is merely a smoke screen for a hidden agenda which seeks a reversal of safeguards for workers and consumers that were enacted only after lengthy debate and often after the industry affected was able to limit the regulations or abort their enforcement.

Nobody, surely, would quarrel with the fact that these regulations are often burdensome, that some rest upon imperfect knowledge, and that all are the result of scientific conclusions that may have to be changed as new knowledge is gained. In deciding to intervene, one can err by thinking that nothing is wrong when in fact something is, or by thinking that something is wrong when in fact nothing is. These errors are known to statisticians as Type II and Type I respectively and it is important to note that they are also called consumer's risk and producer's risk. Unfortunately, a decrease in one type increases the probability of the other. It is, however, significant that when statistical controls are applied to costs, Type II errors are regarded as more serious because if a company has a financial hemorrhage that goes unnoticed, it will go bankrupt.[27] If we are that perspicacious about money, should we be any less about our health and wellbeing?

The origin of totalitarianism lies in the desire of rulers to argue with reality in favor of their own ideological preconceptions and self-deceptions, and in the desire of the peoples they govern to be rid of the responsibilities of freedom—a process that Erich Fromm described in 1941 in his *Escape from Freedom.*[28] One would wish it otherwise, but the danger of precisely such miserable changes is ever present in our society, as it was in those in which, with dreadful regularity in this century, free inquiry has been snuffed out. Tacitus[29] again has a message for us: He reminds us that "rare and happy is the time when one can think what one wants and say what one thinks."

[**Note added in proof**—February 1983:

It would have been tempting to revise the foregoing article in the light of events since December 1980. However, the writer has chosen to let it stand exactly as presented because it accurately forecast the scientific concerns that are still very much in evidence.

Governmental support of science has, as expected, taken on a strong militaristic orientation, as evidenced by what is being supported and what has been cut back. Reindustrialization, starved of funds and technical talent, expired in the deepening recession.

Strong efforts to establish creationism as a "science" ended in disaster in the courts for its proponents. However, quasi-religious obscurantism is not dead; rather, there was evidence of successful anti-evolutionist pressures on the publishers of school textbooks in the sciences.

In the larger public arena it did not prove possible to repeal the Clean Air Act and other anti-pollution legislation. The environmental movement showed great political strength and its views were widely accepted by the public.[30] However, the Executive Branch, through its political appointees, showed itself adept at changing the point and emphasis of regulations as well as their enforcement, which, thanks to budget cuts and political pressures, borders on nonfeasance in some areas.[31] In early 1983, Congress and the Executive were in serious Constitutional conflict regarding the enforcement of regulations for the disposal of hazardous wastes.

Finally, while it is true that the worst conceivable developments were avoided, there has been no significant progress in the solution of the relevant scientific problems well defined in 1980. Everyone's energies were fully taken up with protecting what had been so laboriously achieved. The controversies in prospect at the beginning of the Reagan Administration therefore seem certain to remain an important part of the political scene in the remainder of his term.]

REFERENCES

1. CENTER FOR THE STUDY OF AMERICAN BUSINESS. 1979. Overregulation: Can We Afford It? Exxon USA, Second Quarter, p. 12.
2. SINCLAIR, U. 1951. (1905) The Jungle. Harper & Brothers. New York. p. 65.
3. Consumer Reports. 1979. Hair Dyes. August, p. 456.
4. SINCLAIR, U,[2] p. 117.
5. 1975. Fruit/Vegetable Pulp Extender Replaces Up to One-third Fruit Solids—Up to 50 Percent Coconut in Confectionery. Food Processing, October, p. 100.
6. KARPF, A. 1980. Banned Drug Dumped in Mozambique. New Statesman, August **15:** 2.
7. INSIGHT TEAM of the *Sunday Times* of London. 1979. Suffer the children: The story of Thalidomide. Viking. New York. pp. 64–85.
8. SINCLAIR, U.,[2] pp. 98–99.
9. O'TOOLE, J. *et al.* 1973. Work in America. MIT Press. Cambridge, MA. pp. 76–92.
10. UNITED STATES DEPARTMENT OF LABOR, OCCUPATIONAL SAFETY AND HEALTH ADMINISTRATION. 1980. Access to Employee Exposure and Medical Records. 29 OFR Part 1910, Federal Register, May **23:** 35212–35303.
11. ULLMANN, J. E. & S. E. GLUCK. 1980. Manufacturing Management: An Overview, 2nd. edit. Transwin Publishers. Columbus, OH. pp. 153–157.
12. TACITUS. 1854. Works of Tacitus. Agricola, 30. Henry G. Bohn. London. p. 30.
13. 1976. Water: Arizona's most precious resource. Arizona Business, August–September, p. 26.
14. KING, S. S. 1980. Block Sees Food as U.S. Weapon in Foreign Policy. *New York Times,* December 24, p. 1.
15. CRITTENDEN, A. 1980. Soil Erosion Threatens U.S. Farms' Output. *New York Times,* October **26:** p. 1.
16. BLAU, P. W. & M. W. MEYER. 1971. Bureaucracy in Modern Society. Random House. New York. p. 28.
17. TACITUS.[12] Annales, III, 27.
18. ULLMANN, J. E. 1980. See What You Made Me Do. In Private Management and Public Policy. L. Benton, Ed. Lexington Books. Lexington, MA. pp. 189–206.
19. ULLMANN, J. E. 1978. Tides and Shallows. *In* Management for the Future. L. Benton, Ed. McGraw-Hill. New York. pp. 283–298.
20. 1980. A strategy for rebuilding the economy. Business Week, June 30, p. 146.
21. 1980. Harsh necessity . . . Business Week. September 15, p. 162.
22. MENCKEN, H. L. 1955 (1925). The Hills of Zion, In Memoriam: W.J.B. *In* The Vintage Mencken. Vintage Books. New York. pp. 153–167.
23. GARDNER, M. 1957. Fads and Fallacies in the Name of Science. 133. Dover Publications. New York. p. 133.
24. WHITE, A. D. 1955 (1895). A History of the Warfare of Science and Theology of Christendom. George Braziller. New York. *passim.*

25. GARDNER, M.[23] pp. 337–339.
26. FRUCHTBAUM, H., Ed. 1973. The Social Responsibilities of Engineers. Ann. N. Y. Acad. Sci. **196** (10): 409–473.
27. ULLMANN, J. E. 1976. Quantitative Methods in Management. McGraw-Hill. New York. pp. 166–8, 253–4.
28. FROMM, E. 1965. Escape from Freedom. Avon Books. New York. *passim.*
29. TACITUS,[12] Historiae, I, 1.
30. 1983. A Call for Tougher, not Weaker Antipollution. Business Week, January 24, p. 87
31. ULLMANN, J. E., Ed. 1983. Social Costs in Modern Society. Greenwood Press. Westport, CT. *passim.*

Quid Pro Quo for Public Dough[a]

PETER RAVEN-HANSEN

George Washington University
National Law Center
Washington, D.C. 20052

[At the Section's request, Professor Raven-Hansen first briefly describes a recent case in which the Supreme Court held that a scientist does not become a public figure for purposes of defamation law merely because he accepts public research funds. In contrast, Professor Raven-Hansen argues in the remainder of the paper, such a scientist does assume at least a "quasi-public" status in terms of his accountability for the public research funds and his obligation to disclose his data to the funding agency and sometimes to the public.]

INTRODUCTION

In 1979, to the delight of the research scientists who followed the case, the Supreme Court decided that a research scientist whose work is ridiculed in a "Golden Fleece Award" for wasteful government spending may sue the grantor, a United States Senator, for defamation under the applicable state law. The Court rejected the argument that a scientist, by seeking and accepting public research funds, becomes at law a "public figure" who bears a heavier burden of proof of defamation than other private citizens.

But both lower courts who heard the case had ruled that public funding *does* carry with it a special status for the recipient. While this view was rejected by the Supreme Court in the narrow context of defamation law, it is still valid in other contexts. For example, scientists who accept public research funding do assume a "quasi-public" status in that they must (1) account for their work effort and hours, (2) give the funding agency access to research data and results in accordance with the research grant, and (3) accept the possibility that such data and results may be disclosed to the public at large, including competitive researchers. Each of these examples is explored below, after a brief description of the "Golden Fleece" case and the conflict of attitudes which it suggested.

THE GOLDEN FLEECE CASE

The plaintiff who brought the "Golden Fleece Case," *Hutchinson v. Proxmire,*[1] was Dr. Ronald Hutchinson. He is described by the Supreme Court as a behavioral scientist who specializes in the study of animal and human aggression. Until 1975, Dr. Hutchinson was employed by a public

[a]This paper was presented to the Section of Science and Public Policy of the New York Academy of Sciences on March 4, 1981.

0077–8923/83/0403–0083 $0.175/2 © 1983, NYAS

university and by a state hospital; since that date he has been president of his own non-profit "Foundation for Behavorial Research." Between 1966 and 1975, he directed a series of behavioral research projects funded by the National Aeronautics and Space Administration (NASA), the National Institute of Mental Health (NIMH), the National Institute on Drug Abuse (NIDA), the National Science Foundation (NSF), and the Office of Naval Research (ONR), with grants variously reported as totaling between $820,000 and $910,500. Many of these projects were apparently in the same area: the study of displays of aggression in animals, including biting and jaw clenching by angry monkeys.

The defendant in the case was William Proxmire, the senior Democratic Senator from Wisconsin. Senator Proxmire has served for some time on the subcommittee of the Senate Committee on Appropriations responsible for budget oversight of the agencies that funded Dr. Hutchinson. He has developed a reputation as an untiring critic of wasteful government spending. In March 1975, Senator Proxmire initiated the Golden Fleece of the Month Award to publicize egregious examples of wasteful government spending. The Golden Fleece Award was an immediate media success, as the initial awards were widely reported in the press. Early award recipients included NIH (for what was described as a $102,000 study of whether sunfish grew more belligerent from drinking gin or from drinking tequila)[2]; the Law Enforcement Assistance Administration (for a $27,000 study of why prison inmates like to escape); NSF (for an $84,000 study of why people fall in love); and most recently, the Presidential Inauguration Committee (for charging the taxpayers with a $4,000,000 bill for 1120 around-the-clock military chauffeurs for some 274 inaugural guests and officials, such as President Reagan's personal housekeeper).[3]

In 1975, a member of Senator Proxmire's staff read an abstract of one of Dr. Hutchinson's studies and then spent forty hours following up this lead by calls to the funding agencies. The staff concluded that the research was "obvious" and duplicatively funded, in that several government agencies appeared to be paying for the same work, and one staff member so advised NSF. A Golden Fleece Award was accordingly prepared for NSF, NASA, and ONR, and Dr. Hutchinson was notified by Senator Proxmire's staff. Over Dr. Hutchinson's protest that the Award was inaccurate and unfair, Senator Proxmire announced it by a public speech printed in the *Congressional Record,* by press releases to some 275 media outlets, by mention on the Mike Douglas show, and finally, by newsletter to some 100,000 of the Senator's constituents.

A short quotation from the Senator's speech gives a flavor of the announcement:

> The funding of this nonsense makes me almost angry enough to scream and kick or even clench my jaws. It seems to me it is outrageous.
>
> Dr. Hutchinson's studies should make the taxpayers as well as his monkeys grind their teeth. In fact, the good doctor has made a fortune from his monkeys and in the process made a monkey out of the American taxpayer.
>
> It is time for the federal government to get out of this "monkey business." In

> view of the transparent worthlessness of Hutchinson's study of jaw-grinding and biting by angry or hard-drinking monkeys, it is time we put a stop to the bite Hutchinson and the bureaucrats who fund him have been taking out of the taxpayer.[4]

Although Senator Proxmire's staff followed up the Award with phone calls to the funding agencies, and his newsletter to constituents later claimed that the criticized studies had been dropped, Dr. Hutchinson, in fact, subsequently received at least three more grants from NASA, NIDA, and NIMH totaling over $90,000.

Nevertheless, he mounted a counteroffensive. He first issued his own press release denying the allegations contained in the Golden Fleece Award. Then, in April 1976, he filed a libel and slander suit against the Senator and his staff in federal district court asking $8,000,000 in damages for loss of respect in the profession, injury to his feelings, humiliation, public scorn, extreme mental anguish, physical illness and pain, and loss of income.

Senator Proxmire raised two basic defenses to this lawsuit. First, he claimed that his Golden Fleece Awards were protected by the Speech and Debate Clause of the Constitution, which states that no member of the Congress shall be questioned in any other place but Congress for any speech or debate in either house.[5] In essence, this defense is an argument that the Senator is immune to suit on account of the Golden Fleece awards because they are part of the legislative process which the Constitution protects. The trial court agreed in part, holding that the original speech announcing the Award, the antecedent staff investigation and the press release were all within the investigative and "informing" functions of Congress and therefore protected by the Speech and Debate Clause. However, the court did not agree that statements made on the Mike Douglas show or in the newsletter to constituents were equally protected.

As to these statements, the Senator had to rely on a second defense—a First Amendment defense to libel or defamation actions that is sometimes called "the public figure defense." In 1964, the Supreme Court held that public officials who sue a publisher for defamation have to prove actual malice—that is, bad faith or evil intent—in order to recover damages.[6] In other words, merely erroneous statements by a publisher are insufficient to support an award of damages to a public official. The theory of this decision was that any lesser standard of proof would discourage healthy public criticism of persons in power.

Eventually the rule was extended to so-called "public figures," persons who inject themselves into an area of public contoversy by reason of the notoriety of their achievements or their deliberate courtship of publicity.[7] For example, convicted murderer Son of Sam could certainly be viewed as a public figure both because of his notorious criminal acts and because of interviews that he voluntarily gave to the press. Senator Proxmire reasoned that Dr. Hutchinson had injected himself into a controversy about the merits of the public funding of his research by applying for and accepting such funding and by his press releases and publications about his work.

The trial court agreed with this analysis. Accordingly, Dr. Hutchinson was required to prove malice or evil intent on Senator Proxmire's part in awarding the Golden Fleece. But there was simply no evidence in the case record that the Award was anything more than negligent or erroneous, if that. As a result, the trial court entered a judgment for Senator Proxmire. Dr. Hutchinson promptly appealed this decision.

The Court of Appeals substantially upheld this result. Although it defined Speech and Debate immunity more narrowly than had the trial court, it found that the public figure defense applied to all statements by Senator Proxmire that were not otherwise immune. It noted especially that Dr. Hutchinson was "not forced to seek public funds," and that he had himself generated publicity in this matter by his "public affirmation of the soundness of his research and continued public funding thereof."[8]

Undaunted, Dr. Hutchinson appealed to the Supreme Court.

At this point, the American Association for the Advancement of Science and its affiliate, the American Psychological Association, were stirred to action. They filed a well-drafted and restrained "friend of the Court" brief in the Supreme Court taking issue with the public figure defense that had been upheld by the lower courts. Noting that some 72% of all university-based science research is publicly funded, the associations argued that application for public funding could hardly be considered an effort to engage the public's attention, but was instead merely an initial step toward a "contribution to a discrete body of scientific knowledge." Moreover, subsequent publication of the research results was informational and not advocatory.[9]

The Supreme Court agreed. It needed little space to conclude that Dr. Hutchinson had not become a public figure by accepting public funding or publishing his research. These activities, the Court reasoned, "are much like those of countless members of his profession" and do not thrust him or his views into public controversy to influence others. If they made him a public figure, "everyone who received or benefited from the myriad public grants for research could be classified as a public figure," a conclusion that the Court squarely rejected.[10]

As to the Speech and Debate immunity, the Court was prepared to extend it to Senator Proxmire's comments in the *Congressional Record,* but there it drew the line. Republication of those comments by press release or newsletter, no matter how routine or essential such press releases and newsletters are to congressional communication, are not part of the legislative function or deliberative process which the Constitution protects.

In short, Dr. Hutchinson won on all counts in the Supreme Court, and it promptly sent the case back to the lower courts for trial. The effect of the decision was to allow Dr. Hutchinson to prosecute his case without having to prove that Senator Proxmire had acted maliciously or with evil intent, although the Senator was still free to raise other defenses, such as the truth of statements he made in the Golden Fleece Award. But before the the case went to trial, the litigants agreed on a settlement. Senator Proxmire paid Dr. Hutchinson $10,000 out of his own funds[11]—an amount which, at normal hourly rates, would barely cover the plaintiff's attorneys fees—and the lawsuit was dropped. Lest anyone think that the Senator himself should get the next Golden Fleece Award for using the taxpayers' money for his legal defense, he has reportedly assigned the royalties from his book, *The Fleecing of America,* to repay the Treasury for his attorney's fees.[12]

★ ★ ★

The Golden Fleece case is of passing interest to students of libel law and the First Amendment, and perhaps of more than passing interest to those few members of Congress who keep track of the deterioration of their legislative immunity. However, on its face, the decision is at best footnoteworthy as a matter of science and public policy. What is of greater interest is the stark conflict of attitudes which the case exposed between publicly funded researchers on the one hand, and those who pay the funds, on the other.

The attitude of the latter was suggested by the decisions of the lower court judges in this case. Those decisions implicitly reflect the view that the researcher who accepts public funds (hereinafter called "grantee") incurs special obligations or duties—a special status, in effect—which is different from the status of those who are not funded by the federal government. The Supreme Court refused to accept this view for purposes of libel law only; it never reached the broader question of what public responsibilities and status go with public funds.

In contrast, the attitude of the grantees is that they do not incur special public responsibilities or a public status simply by accepting public funds. They view themselves as free and independent researchers whom the public is privileged to support, although few go so far as to claim that public funding is a God-given right. They assert that the only responsibilities which they owe are those responsibilities and duties owed by all researchers or scientists as professionals and that any insistance by the government or the courts on a special status for grantees is not only mistaken and misguided, but a direct assault on scientific freedom.

Who is right? In this era of clichés about getting the government off our backs and in this particular forum, the sentiment probably lies with the grantees. But this paper takes the opposing view: grantees *do* incur special responsibilities and *do* assume a quasi-public status for some purposes for at least the duration of their grant. This is a *quid pro quo* for public dough. Moreover, I would state it more forcefully: many of the grantees' views to the contrary are not just wrong, but naive, disingenuous, arrogant and, in some ways, harmful to the very principle they invoke—scientific freedom.

Let me illustrate the point by taking three examples that have aroused the ire of grantees.

100% Effort Reporting

Example 1: OMB Circular A-21 establishes accounting rules for research grants to educational institutions. The underlying principle of the Circular is that a cost is allocable to a grant only if it is incurred solely to advance the work intended under the grant, benefits the grant effort in some determinable proportion, or is necessary to the sponsoring institution and properly assignable to the grant. To implement this principle, the Circular originally required individual researchers to prepare activity reports annually or each semester allocating 100% of their work time among the funded research, other kinds of research, instruction and other "major functions" of the sponsoring institution.[13]

It is hard to think of any federal regulation concerning research (except perhaps for regulations concerning Institutional Review Boards) that has generated more passion and annoyance in the scientific community. The National Academy of Sciences, for example, after what science reporter Daniel Greenberg characterized as a "foam-at-the-mouth session,"[14] issued a statement that Circular A-21

> would further constrain the already limited flexibility in research thrust, increase the administrative burden, reduce morale among teaching and research personnel, and provide a cumbersome meaningless documentation in terms of percent-of-effort for a continuum of scholarly activities.[15]

Yale President, G. M. Giamatti, was more colorful, in keeping with his humanities background, when he reportedly said:

> Never have I seen the lash of federal regulation applied to a crucial area of the nation's intellectual life with such seeming indifference to financial and human consequences.[16]

Even one of the less rhetorical critiques appearing in *Science* charges that the 100% effort reporting requirement is an "improper invasion of university prerogatives" that creates "constraints counter to the spirit and intent" of science.[17]

A mathematician, writing to *Science* to criticize an earlier effort reporting proposal by the federal government, was less restrained. He said that "short term scientific accounting is an absurdity . . . noxious to scientific performance and inhibiting to scientific thought."[18] He then concluded that "the percentage information concerning distribution of 'efforts' in various categories is none of the university's, the government's or accountants' business." Instead, once a grant is made, "it should be treated like a salary, that is, irrevocable," and the most that can be asked is an occasional "brief report of the year's progress. . . ."[19]

At least some of these protests badly miss the mark and undervalue the strong public interest in assuring accountability for public funds committed to research.[20] Although the still-accepted cliché is that grants are to support and not to procure, the grantee must use the funds solely for the grant purposes in accordance with the bookkeeping and auditing requirements set by the government.[21] Moreover, precisely because grant money is used to support basic research, the "success" or benefit of which is difficult to judge, accountability necessarily focuses on the mechanics of fund expenditure rather than scientific results. As former Comptroller-General Staats said, because substantive accountability or quality control has effectively been "taken out of the public domain" by the peer review system, it "stands to reason that the public requires increased *fiscal* accountability for university research in order to retain some check on public research expenditures."[22] As a result, by virtue of public funding, a researcher's time takes on a public aspect, and he incurs a responsibility to account for it.

One argument against this assessment is that there is no need for controls like A-21 because there is little or no fraud or misuse of public funds by grantees:[23] "If the faucet ain't leaking, don't fix it." For example, the General Accounting Office recently found only 0.23 percent of $1.5 billion in grant funds to have been improperly charged.[24]

Yet the Department of Health and Human Services, a major funding agency, recently issued new regulations governing the debarment and suspension of individuals and institutions from financial assistance for fraud and abuse of public funds, asserting that they were "necessary because there have been instances of fraud or abuse by persons who are responsible for Federal funds."[25] The comments accompanying the regulations explain that serious cases of misappropriation of grant funds, willful failure to comply with record retention and access requirements, and intentional falsification of reports prompted the new regulations. In a similar vein, the Washington *Post* recently reported what it characterized as a "crime wave" in scientific research, principally involving the faking of data.[26] If there *is* an upsurge of falsification in the reporting of results, the same character flaws and pressures which cause this kind of fraud may lead to other kinds, including fraud in the administration of grant funds.

Moreover, the "faucet ain't leaking" argument rests on an unduly narrow concept of grant abuse. The principal public concern underlying Circular A-21 is not with fraud and abuse in a criminal or moral sense, but with the diversion of grant funds from their intended use by sloppy or indifferent management and "innocent" manipulation of the funds by grantees.[27] The same GAO report cited above reported that almost 6% of the studied grant expenditures had to be set aside as "not auditable," presumably due to inept bookkeeping.[28] Everyone who has ever worked in a university knows that grantsmanship often involves not merely the getting of grants, but their manipulation to pay for research or clerical assistance on non-grant matters or matters for which funding is not directly available. Even a severe and responsible critic of Circular A-21 concedes no less than seven abuses of the grant system, of which the largest is undoubtedly the improper transfer of funds between grants.[29] In fact, as availability of research funds declines, pressure will increase to transfer funds between grants and to obtain multiple grants for essentially the same work, in order to support the kinds of research that are not funded.

Another abuse of public funding is associated with the deteriorating financial position of research-sponsoring universities, which has caused them to overburden their public grants with indirect costs. Indirect costs are said to have risen from 15%–20% of institutional salaries and wages to 50%–70%, in part to make up for the effects of inflation on institutional endowments and other cost factors.[30] Science reporter Greenberg estimates that, as a result, perhaps as much as one-half of the federal appropriation for science research never gets to the laboratory.[31]

While all of these forms of grant manipulation may be rational and defensible from the viewpoint of science alone, they are clearly contrary to the congressional intent that public funds be spent on specific projects for specified purposes. Indeed, that intent was instrumental in the promulgation of Circular A-21 itself; the preamble to the circular explains that it was developed at the "urging" of the House and Senate Appropriations Committee.[32] Thus,

> [s]cientists cannot simply take refuge in the assertion that a university is an honest place, or that all the money transferred is somehow used for science. Accounting is necessary for cost transfers and careful records are needed for salaries charged to government-sponsored projects.[33]

Contrary to the criticism of A-21 noted above, it *is* emphatically the government's and its accountants' business what and how much is charged to publicly funded research grants.

But a second and stronger argument against fiscal controls like Circular A-21 is that diversion of public funds is *necessary* to the effective conduct of basic research. Of course, this argument is not made this bluntly, but instead is phrased in terms of the meaninglessness, impracticality, inaccuracy or inevitable untruthfulness of effort reporting, given the "continuous nature of research inquiries."[34] The argument goes that scientists are independent thinkers, not professionals employed to do a job, and their thought processes and research efforts therefore cannot be subdivided by individual projects or functions; the process pervades every aspect of their academic work. Thus, scientists do work which is too complex for 100% effort reporting, unlike lawyers, who supposedly can report their time in this fashion "because their tasks are of uniform character (always advising a client). . . ."[35] Since public funding in reality supports a process, not a project, it is idle to worry excessively about which project grant is charged for the researcher's time.

This argument is made convincingly by a number of articulate scientists, but let me play devil's advocate. Is an effort reporting requirement as impossible and impractical as its critics assert? Lawyers keep track of their time on a daily, hourly, or even quarter-hourly basis, and contrary to the comment quoted above, their tasks are *not* of uniform character. Why is the allocation of the time a scientist spends instructing a graduate student significantly more difficult than the allocation of the time a senior law partner spends instructing a young associate on how to prepare a court pleading? Both involve the allocation of time among instruction, administration and client/grant activity. Of course, complete accuracy is impossible, but Circular A-21 does not demand complete accuracy. It takes pains to recognize that

> because of the nature of work involved in academic institutions, the various and often interrelated activities of the professorial and professional employees frequently cannot be measured with a high degree of precision, . . . reliance must be placed on reasonably accurate approximations, and . . . acceptance of a degree of tolerance in measurement is appropriate.[36]

One cannot push this argument too far, because, of course, the practice of law and basic scientific research *are* dissimilar, but are not critics of effort reporting overstating their case?

More critically, it is *their* burden to come up with an alternative to Circular A-21 to assure fiscal accountability for public grant funds. However rational, or even essential, it may be to fund a "continuum of research activities," this is not the road that Congress has taken. We generally fund research on a project-by-project basis; we have yet to enact a negative income tax for research scientists. Accordingly, some system for tracking an accounting trail to each project is appropriate. The proper objection to Circular A-21 is that its tracking is more disruptive of the academic workplace than it needs to be. But if this true, then the critics should suggest less disruptive alternatives that will achieve the same purposes.

One such alternative is the proposal that grantees "self-certify" the reasonableness and fairness of their charges.[37] But if grantees will deliberately fudge 100% effort reporting, as its critics contend,[38] and if indifferent management is at the root of the perceived problem, how can the government rely on self-certification by indifferent researchers to meet its needs? Another alternative that has been suggested is to delegate the entire fiscal monitoring task to the academic institutions that employ the researchers.[39]Yet it is the universities' overloading of their indirect cost charges to federal grants which in part prompted the issuance of Circular A-21. Perhaps the answer is to combine these proposals with spot audits conducted by government auditors who are specially sensitized to the nature of the research process in an academic setting, as the Sloan Commission on Government and Higher Education has reportedly urged,[40] but this suggestion begs the question of what data the auditors should examine.

I have no answer, and in fairness to the quoted critics of Circular A-21, I should note that most of them now acknowledge the need for some appropriate system of accountability. The point is simply that the issue is confused by rhetorical cries about the "invasion" of scientific freedom by the government, government prying into the private "business" of the independent scientist and the like. The researcher who voluntarily seeks and accepts federal grants is no longer the independent loner of science lore; he becomes *part* of the public business to the extent he uses public funds. The task, which falls equally on the shoulders of the government and grantees, is to devise the least invasive means of accounting for this aspect of the public business.

Agency Access to Data

Example 2: A social scientist receives a grant from the Department of Health and Human Services to evaluate the operation of a government agency. In conducting his research, he interviews a number of agency employees, generating raw questionnaire data and a matching list by which interviewees could be identified. Unknown to the funding agency, the researcher makes pledges of confidentiality to the interviewees at the commencement of his survey. Subsequently, the funding agency asserts the right to examine the raw data and the matching lists, partly to assure that funds were properly spent, and partly to make the data available to future grantees or contractors continuing this line of research, thus avoiding duplication of completed work. The researcher resists this request on the ground that it violates the pledges of confidentiality he has made.

This example is loosely drawn from a case brought to the attention of the American Association for Advancement of Science.[41] The researcher's position is quite clear: confidentiality "assurances are standard practice for responsible social science research," and if there was any reason why this standard practice could not be followed, "it should have been specified by the agency project officer."[42] He further contends that there was no limitation on this practice in the grant conditions, and the funding agency never gave a valid reason for requiring the matching list. In short, the government demanded a breach of confidentiality in disregard of social science practice and of the risk to effective research in the future.

Once again, this critique misses the mark. There is no question about the importance of protecting confidentiality or about the government's obligation in an appropriate case to honor assurances of confidentiality.[43] But there is another public interest at work here: a public agency's need for access to data to evaluate, validate and audit the results of research that it funds. However free scientists are to give pledges of confidentiality when they are "working on their own nickel," they may be constrained when they accept public funding. They must then expect and plan for agency access to the resulting research data.

This is not to say that an agreement to assure confidentiality cannot be negotiated with the agency ahead of time. But it is the researcher's responsibility to obtain it, as an exception to the rule of agency access to publicly funded research. Agency access to data is a *quid pro quo* of public funding. If it needs to be modified owing to specific research needs, let the person who knows them best—the researcher—come forward and ask for the modification in advance.

Moreover, this rule, or presumption, or *quid pro quo,* is not simply an unwritten incident of public funding. Ordinarily it is reduced to writing in the grant boilerplate. In the very example cited, the written grant conditions included one entitled "Reports, Records, and Inspections." This condition not only compels periodic reporting, record maintenance, and on-site access by agency representatives, but expressly requires the grantee

> to furnish such information as in the judgement of the [agency] representatives, may be relevant to a question of compliance . . . or the effectiveness, legality, and achievements of the program.[44]

Similar access or "rights-in-data clauses" appear in many, though not all, federal research grants and put the researcher on notice of his responsibility to seek an exception or modification when one is dictated by research needs.[45] Since grantees have perforce mastered the government's paperwork sufficiently to land the grants, it is not unreasonable to expect them to read and comprehend the grant conditions as well.

Here again, public funding creates a difference between what an agency official in this instance called "the right of *private* researchers to maintain the pledges of confidence which they may make," and the right of federally funded researchers.[46] In this respect, too, grantees assume a public responsibility or obligation that sets them apart from scientists at large.

Public Access to Data

Example 3: A group of scientists receives a federal research grant from a branch of the National Institutes of Health to conduct a long-term study of diabetes patients. Primary data are generated by the group over several years as they prepare to publish their conclusions. But then the funding agency receives two requests for the data under the Freedom of Information Act. One request is by a scientist who wants to use the data to publish his own article; the second is by a large drug company that hopes to rebut certain conclusions which it fears the grantee will draw from the study. Approval of the requests will give the requesting scientist the chance to publish before the researchers who actually compiled the data and may subject the study to attack by the drug company before the grantee has completed validation of the data to its satisfaction.

This example is a composite of several reported court cases involving Freedom of Information Act (FOIA) requests for publicly funded research data.[47] Release of such information has been vigorously opposed by grantees on three main grounds. *First,* there is what one group of social scientists has candidly described as the "obvious need for first rights to mine the data . . . , the importance of a temporary monopoly or type of patent."[48] Others have called this the researcher's "stock-in-trade"; Philip Handler of the National Academy of Sciences has called it the "property" of scientists[49]; a prominent biologist has called it a researcher's "moral right" to first interpretation of data that is "generally accepted [as an] underlying postulate of scientific conduct. . . ."[50] *Second,* a related ground for resisting disclosure is that researchers might lose potential patent rights.[51] Finally, some opponents of disclosure have cited the danger of misinterpretation or abuse of prematurely disclosed data,[52] or, remarkably, the risk of discovery of error.[53]

These arguments misapprehend (or chose to ignore) the quasi-public status of the grantee.[54] The FOIA creates a presumption in favor of disclosure of information contained in agency records, unless records are expressly exempted by statute.[55] While the Supreme Court has ruled that data retained by the grantee, and therefore not in the physical possession of the agency, are not ordinarily "agency records" subject to the FOIA, the implication of the decision is that grantee-developed data held by the agency for some agency purpose *are* subject to the FOIA.[56] In fact, so strong is the FOIA policy that a federal court recently ruled that data on breast-feeding practices which had been sent by a private research group for analysis to the Center for Disease Control (CDC) had to be released to two baby formula manufacturers under the FOIA even though the CDC had not funded the data collection and was, in effect, only gratuitously lending its analytic facilities to the private research group.[57]

Thus, if the diabetes study data in the example are in the possession of the agency for an agency purpose, any so-called proprietary interest of the scientist in the data gives way to the public interest in disclosure. As the D.C. Circuit Court of Appeals puts it:

Scientists engaged in research on federal grants must accept the fact that any documents filed with the federal government, whether on the scientist's own

> initiative or an audit or other lawful demand, are subject to FOIA. Even in scientific terms, any such audit provides a surrogate for the kind of reliability usually accorded to scientific studies by replication of experiments when feasible.[58]

Usually the very purpose of public funding is to encourage the earliest possible exposure and exploitation of the research for the public good. Why should this purpose be subordinated to the private proprietary interest of a publicly funded researcher? Moreover, given the headstart that the grantees will ordinarily have in the understanding of their own data, one may doubt the need for any delay in public disclosure to let them "capitalize on [their] investment," as some social scientists put it.[59] If, as one court said only partly in jest, researchers really are "a mean spirited lot who pursue self-interest as ruthlessly as the Barbary pirates did in their own chosen field,"[60] they will probably manage to "scoop" the FOIA requester most of the time.

Protection of potential patent rights seems an even weaker ground for nondisclosure of grant data. Many grant proposals contain conditions reserving patent rights in the government, at its option. When this is the case, it is unlikely that researchers with potentially patentable concepts would risk submitting them to the government in the first place. But if they do, why should their personal financial interest be preferred over anyone else's interest when the public paid the bills?

Finally, the danger of misinterpretation or abuse of the data also seems a dubious ground for nondisclosure. Ironically, the National Commission for Protection of Human Subjects of Biomedical and Behavioral Research found that

> reported instances of distorted interpretation of hypotheses and data appear to be generally attributable to material released to the press by investigators on their own initiative.[61]

Although strong arguments have been made for postponing release of preliminary epidemiological studies, clinical trial data and other primary data sets until they have been completed and verified,[62] until such an FOIA exemption is enacted, we must rely on critical review in the scientific media to guard against misuse of such data. But that has always been the scientific community's answer to bad research. Why should it not suffice in the example described? As for the risk of discovery of error, if this is a *danger,* "it is a sad day for the scientific community and the FOIA."[63]

Moreover, there is another basis for public access that does not turn on physical possession of the data or the FOIA. If, in promulgating a regulation, an agency relies upon a grantee's data, whatever its location or state of completion, then the agency may be obliged to disclose that data in response to a legal challenge to the regulation. Under the Administrative Procedure Act, as interpreted by the courts, an agency may have to include the data in the administrative record produced to support the regulation[64] and may even be obliged to disclose them for public comment *before* the regulation is promulgated.[65] Thus, in the same case in which the Supreme Court ultimately denied FOIA access to data retained by the grantee, the Food and Drug Administration had to disclose some of the data to support a regulation based in part upon them.[66] If the agency does not possess the data, it may be forced to demand access to them under the conditions of the grant, or lose the legal challenge to its regulation.

In my view, doubts about FOIA disclosure of research data disappear in these circumstances. Here the data are sought to probe the basis of administrative action, and if they are good enough for the agency to "use," they are certainly sufficiently complete for disclosure. Disclosure of the research data in these circumstances helps to assure good agency decision-making (notably in the areas of health, environment and safety), compared to which the researcher's "stock-in-trade" rights fade into insignificance. The dissenting observation in the Supreme Court FOIA case cited above is especially *apropos* in these circumstances:

> . . . [T]he understandable tendency of agencies to rely on non-governmental grantees to perform myriad projects distances the electorate from important information by one more step. If the records of such an organization, *when drawn directly into the regulatory process,* are immune from public inspection, then government by secrecy must surely return.[67]

CONCLUSION

This paper has explored the proposition that researchers who seek and accept public funding subject themselves to responsibilities and constraints which do not affect those who are not so funded. Their work becomes in important respects public business, (1) for which they must account in annoying detail and (2) to which they must allow access by the agency and (3) often by the public. In this sense, they *do* assume a quasi-public status, like it or not. A wooden and sometimes melodramatic resistance to these demands of open and accountable government is misplaced and counterproductive. The scientific community should instead acknowledge their legitimacy, but strive to minimize their invasiveness.

So how does all this relate to the *Golden Fleece Case?* Maybe not very directly. But consider the case if the arguments I have attributed to the grantees should prevail in the three cited examples. Then the government could not ask for effort reporting; all that one critic will agree to is an annual "brief report of the year's progress. . . ." The funding agency could be blocked from access to the survey data it funds whenever the grantee unilaterally follows what one describes as the "standard practice" of pledging confidentiality to interviewees. Nor could the public access the data under the FOIA, at least until the grantee was good and ready to publish or transmit them to the funding agency.

In these circumstances, what fiscal control is left to the government? One answer that comes to mind—maybe the most important—is congressional budget oversight. But how is such oversight to be accomplished, given the scarce congressional resources for detailed study, limited attention spans and multiple legislative responsibilities? Clearly it is achieved by using the investigative and hearing process to identify spending abuses; enlisting the media to focus attention on the worst of them; and, thus, by example, forcing the funding agencies themselves to monitor their grants.

That is exactly what Senator Proxmire often achieves by his Golden Fleece Award, and it is a good thing. And what scientists would disagree, if his awards spared the research sciences and went exclusively to boondoggles funded by the Departments of Defense or Transportation (as most have)? Moreover, I submit that occasional inaccuracy or exaggeration, which is inevitable given limited congressional staff resources, is a small price to pay for the public good which comes from the exposure of waste.

Happily, I am not alone among friends and members of the scientific community in drawing this conclusion. After the Supreme Court's "Golden Fleece" decision, the Council of the Federation of American Scientists (FAS) took a strong editorial position in favor of creating a "good faith" defense for legislators charged with defamation for statements like those made by Senator Proxmire. It noted that

> [t]he Legislature needs the right to criticize, in particular, all projects funded by public monies without fear of libel suits if they get something wrong. . . . After all, Congressmen are constantly popping off and are elected to do so.[68]

The FAS Council looked beyond the parochial self-interests of grantees in proposing this defense. All research scientists who seek and accept public funding need to do the same in weighing the "*quid pro quo's*" of public funding.

[**Note added in proof:** The discussion in the text of "100% Effort Reporting" concerns the 1981 version of Office of Management and Budget (OMB) Circular A-21, which established various reporting requirements for recipients of federal research grants. Since this paper was presented in March 1981, OMB solicited public comment on a proposed revision of Circular A-21, 47 Fed. Reg. 932 (1982), and promulgated that revision in 1982. Circular A-21, Cost Principles for Educational Institutions (revision), 47 Fed. Reg. 33658 (1982). In doing so, OMB responded to some of the academic criticisms noted in the text and substantially relaxed its work effort documentation requirements by adopting several alternatives proposed by the Association of American Universities and the Council of Science Society Presidents. *Id.* It is still too early to predict whether revised Circular A-21 will blunt all of the academic criticisms or whether the alternatives that have been adopted will satisfy the public need for accountability for research funds.]

NOTES AND REFERENCES

1. *Hutchinson v. Proxmire,* 443 U.S. 111 (1979), *reversing* 579 F.2d 1027 (7th Cir. 1978), *affirming* 431 F.Supp. 1311 (W.D.Wis. 1977). The facts in the text are drawn from these three judicial opinions.
2. Farney, D. 1979. A Bureaucracy Grows in Congress as Panels and Staff Mushroom, *Wall Street Journal,* Dec. 18, at 1, col. 8.

3. Inauguration Alleged to Cost $14 million in "Hidden" Expenses, *Washington Post,* Feb. 15, 1981, § A at 18, col. 1.
4. 121 *Cong. Rec.* 10803 (1975).
5. U.S. Const. Art. 1, § 6.
6. *New York Times Co. v. Sullivan,* 376 U.S. 254 (1964).
7. See, *e. g., Rosenbloom v. Metromedia, Inc.,* 403 U.S. 29 (1971).
8. *Hutchinson v. Proxmire, supra,* 579 F.2d at 1035 n. 14.
9. Brief of American Psychological Association and American Association for the Advancement of Science as *Amicus Curiae, Hutchinson v. Proxmire, supra,* 443 U.S. 111.
10. *Hutchinson v. Proxmire, supra,* 443 U.S. at 135.
11. Author's communication with office of Senator Proxmire, March 2, 1981.
12. *Id.*
13. OMB Circular A-21, 44 Fed. Reg. 12368 (1979).
14. GREENBERG, D. 1980. Clocking Science, *Washington Post,* Dec. 23, § A at 15, col. 1.
15. *Id.*
16. *Id.*
17. MACLANE, S. 1980. Total Reporting for Scientific Work. Science **210:** 158, 163.
18. Letter from S. LANG. 1967. Science **155:** 773.
19. *Ibid.* In fairness, it should be noted that Professor Lang subsequently softened his views, or at least his rhetoric, and made a well-reasoned and restrained criticism of the 1980 effort reporting requirement. Letter from S. Lang. 1980. Science **207:** 1148.
20. Several billion dollars are expended each year in basic research and university-based research and development. *See generally* American Association for the Advancement of Science, *Research and Development, AAAS Report V* (1980).
21. MASON, M. 1976. Current Trends in Federal Grant Law—Fiscal Year 1976, 35 *Fed. Bar J.* 163, 167 (1976). Even non-funded research can be regulated under the constitutional spending power as long as such regulation is reasonably related to achieving the overall purpose of a grant or grant condition. *See generally* J. Robertson, "The Scientist's Right to Research: A Constitutional Analysis," 5 *S. Cal. L. Rev.* 1203, 1275–77 (1978).
22. STAATS, E. 1979. Federal Research Grants. Science **205:** 18, 19. (emphasis supplied).
23. LANG, *supra,* Science **207:** 1150.
24. WALSH, J. 1980. Universities Face New Accounting Rules. Science **210:** 34, 36.
25. 45 Fed. Reg. 67262 (Oct. 9, 1980).
26. HILTS, P. 1981. Research Results Falsified. *Washington Post,* Feb. 17, § A at 1, col. 1.
27. See, *e.g.,* Report of National Commission on Research. Accountability: Restoring the quality of the partnership. Science **207:** 1177. (hereinafter "Accountability Report").
28. WALSH, *supra,* Science **210:** 36.
29. MACLANE, *supra,* Science **210:** 159.
30. *Id.*
31. GREENBERG, *supra.*
32. OMB Circular A-21, *supra,* 44 Fed. Reg. at 12368.
33. MACLANE, *supra,* Science **210:** 160.

34. See, *e.g.*, *Accountability Report* at 1177; Lang, *supra*, Science **207:** 1149; MacLane, *supra*, Science **210:** 160.
35. *Id.* at 163 n.4.
36. OMB Circular A-21, *supra*, 44 Fed. Reg. at 12375.
37. MacLane, *supra*, Science **210:** 163, *Accountability Report* at 1179.
38. *Cf.* MacLane, *supra*, Science **210:** 160 (fictitious reporting).
39. *Accountability Report* at 1180–81.
40. Grant Regulation; Change Urged. 1980. Nature **284:** 293.
41. *See* D. McTavish, *et al.*, Threats to Confidentiality of Data Sources: a Case Study. (Dec. 26, 1975) (unpublished abstract on file with the author).
42. Letter from D. McTavish to R. Chalk (Dec. 18, 1980).
43. *See* letter from M. Groggin (HHS official) to F. von Hippel (Sept. 17, 1980): "It is incontrovertible that some data could not be accumulated without assurances of confidentiality. . . . We recognize and respect the right of private researchers to maintain the pledges of confidence which they may make."
44. *Id.*, attachment.
45. *Cf. Lack of Guidelines for Federal Contract and Grant Data,* H. Rep. 95-1663, 95th Cong., 2d Sess. 13–14 (1978).
46. Groggin, *supra*, at 1 (emphasis supplied).
47. See, *e.g.*, *Forsham v. Harris*, 445 U.S. 169 (1980); *Washington Research Project v. DHEW*, 504 F.2d 238 (D.C. Cir. 1974); *St. Paul's Benevolent Educ. and Missionary Inst., v. U.S.*, 506 F. Supp. 822, (N.D. Ga. 1980).
48. Commission on National Statistics, *Sharing of Social Science Research Data* (1980).
49. National Commission for Protection of Human Subjects of Biomedical and Behavioral Research, *Disclosure of Research Information Under the FOIA* 20 (1977).
50. Dickson, D. 1980. Research data: Private property or public good? Nature **284:** 292. (quoting Professor J. Edsall). *See generally* R. Merton. 1969. Behavior Patterns of Scientists. *Amer. Scholar* **38:** 197.
51. *See* Wikessen, 1976. Grant application: Panel finds new laws enable stealing of ideas. Science **193:** 301, 302 (focus on data contained in applications for funding).
52. See *Disclosure of Research Information Under the FOIA, supra,* at 10.
53. See *Forsham v. Califano,* 587 F.2d 1128, 1147 (D.C. Cir. 1978) (Bazelon, J., dissenting), *aff'd sub nom. Forsham v. Harris,* 445 U.S. 169 (1980).
54. I would distinguish the *unfunded* grant applicant who has the same proprietary interest in his proposals that an author has in a manuscript submitted for publication. In fact, unfunded grant proposals are apparently not presently being disclosed by funding agencies, although the FOIA does not clearly exempt such documents. *Disclosure of Research Information Under the FOIA, supra,* at 4, 23.
55. See, *e.g.*, S. Rep. No. 813, 89th Cong., 1st Sess. 3 (1965).
56. *Forsham v. Harris, supra,* 445 U.S. 169.
57. See *St. Paul's Benevolent Educ. and Missionary Inst. v. U.S., supra,* 506 F. Supp. 822.
58. *Forsham v. Califano, supra,* 587 F.2d at 1137.
59. Commission on National Statistics, *supra,* at 7.
60. *Washington Research Project v. DHEW, supra,* 504 F.2d at 244.
61. *Disclosure of Research Information Under FOIA, supra,* at 39.
62. Dickson, *supra,* Nature **284:** 292; S. Shapiro. 1979. *Release of Privacy Data Sets from the Producer's Point of View* (1979).

63. *Forsham v. Califano, supra,* 587 F.2d at 1147 (Bazelon, J., dissenting).
64. See *United States v. Nova Scotia Food Products Corp.,* 568 F.2d 240 (2d Cir. 1977).
65. *Id.* at 251, 252:

> We can think of no sound reasons for secrecy or reluctance to expose to public view (with an exception for trade secrets or national security) the ingredients of the deliberative process. . . . When the basis for a proposed rule is a scientific decision, the scientific material which is believed to support the rule should be exposed to the view of interested parties for their comment.

66. *Forsham v. Harris, supra,* 445 U.S. at 177 n.7; *Forsham v. Califano, supra,* 587 F.2d at 1133, 1134 n.12.
67. *Forsham v. Harris, supra,* 445 U.S. at 192 (JJ. Brennan and Marshall, dissenting) (emphasis supplied).
68. Speech and Debate Clause Defended Against Court Decision. *32 FAS Public Interest Report* at 7 (Sept. 1979).

Human Subjects Regulations on the Social Sciences[a]

ITHIEL de SOLA POOL

Department of Political Science
Massachusetts Institute of Technology
Cambridge, Massachusetts 02139

Quietly, without a great deal of fanfare in the media, certainly without much awareness by the general public, the universities in the last years have survived an onslaught fully as dangerous as McCarthyism in the 1950s. Indeed, one could argue that in this century, no attack on academic freedom has threatened its destruction as much as the proposed, but now defunct, IRB regulations of August 14, 1979. On that date the then Department of Health, Education and Welfare published a set of regulations in the Federal Register, under which all of the following things would have been true.

A scholar, like Charles A. Beard, exploring 18th-century records to examine the economic interests of the founding fathers would have had to first get permission from an Institutional Review Board (IRB), since his research might have harmed identifiable individuals. (The regulations did not exclude dead human subjects from protection.)

A visiting scholar on sabbatical, using a university library to study Watergate would have had to ask permission of the local IRB. The regulations covered not only research sponsored by the university, but also all research at the university.

A student writing a paper, would have had to get permission before interviewing a candidate for political office.

None of this was intended, of course.

So it was inevitable that the Department of Health and Human Services (HHS) would modify the regulations somewhat before issuing them, but restrictions close to those ludicrous ones were fully intended.

A social scientist informally interviewing a Soviet dissident would first have to ask him to give explicit informed consent—in disregard of the consequences that would have followed if the Soviet secret police got their hands on such evidence that it was not just a conversation.

A sociologist before observing a demonstration on the street would have to satisfy an IRB that any potential harm that his observations might do to those he observed was justified by the value of his research. If he were a journalist or an ordinary citizen it would have been his right to watch—but not if he were a university scholar doing social research.

On January 22, 1981, a new revised set of regulations was finally issued by

[a]This paper was presented at a meeting of The Section of Science and Public Policy of the New York Academy of Sciences held May 6, 1981.

0077–8923/83/0403–0101 $01.75/2 © 1983, NYAS

HHS, which drew back from almost all of those positions. For one thing, the new regulations apply substantially only to research funded by HHS itself. Even if some of the procedures remain a bit silly, they are simply part of HHS's procedures in deciding how to spend its own (rapidly declining) funds. A scholar without HHS funding can look at the founding fathers, Watergate, or whatever he wishes without IRB permission.

For another thing, a variety of particularly harmless kinds of research, such as most library research and most interviewing, have been exempted from review.

Also, a particularly obnoxious clause that required IRBs to review not only the potential harm to subjects which a researcher might do, but also to evaluate his research methods, has been struck.

In short, the protests of the social science community, the representations of its professional associations, and the advice of lawyers and Congressmen have been heard, and HHS yielded. A victory has been won; so perhaps this is now all history. True, but not quite! There remain a couple of aftershocks that need to be dealt with.

Other departments are now drafting their rules, and some of the same issues rise over and over. The Department of Energy, for example, published proposed rules containing the intolerable clause requiring IRBs to pass on research methods.

More important, there is a major job to be done of educating universities themselves about the proper functions of IRBs. The HHS regulations not only require IRB review of any research that it funds, but also a "general assurance" from the university that any research it conducts, however funded, will be done with proper regard for the rights of human subjects. The university must describe to HHS how it will protect them. A harried administrator, aware that there is an IRB procedure that is used with HHS-funded research is likely to respond, with a letter saying "we will do the same thing with the rest of our research." Then, voluntarily, social research will be back in the same mess all over again. Although HHS has the right to enforce these rules for research that it funds, they are neither appropriate nor intelligent for all research.

The reasons why these rules are improper will, I hope, be made apparent in this presentation, but for the moment let me simply say that the American Council of Education drafted appropriate language that universities can use in their general assurances. They should beware of the temptation to make of HHS's procedures their own voluntary prison.

While some problems thus remain, in general social researchers have won, so I cannot do the easy thing of giving you an agitational speech, calling on you to man the ramparts. Something more thoughtful, more analytic is called for.

Perhaps we should ask: What is there that we have learned from this experience? How could it all have happened? How could it be that in free universities in a free society we came so close to a major debacle, with little awareness of what was going on, and with relative quiescence by students and faculties alike? We can see in these events a pattern that is quite characteristic in the expanding organizations of our ever more institution-ridden and bureaucratized society. It is a pattern of intellectual failure of a new and enlarged populace of decision makers to appreciate the credenda earlier held sacrosanct by a guild whose members had absorbed those canons during their professional training, and had assimilated them into their cadre identity. When institutions pass from being such sectarian enclaves of an idiosyncratic culture to being encompassed into more comprehensive social decision processes, important special perspectives can get lost. That can happen to a newspaper or magazine bought by a conglomerate. It can happen to a university brought under the influence of politics.

Two great principles were at stake in the debate over IRBs. These were (1) anathema to prior restraint and (2) the nature of the university.

PRIOR RESTRAINT

The doctrine of "prior restraint" is a legal concept. To constitutional lawyers it is fundamental. The constitution in Article 1 and the First Amendment draws lines among a set of domains in which Congress is free to act and domains in which it is without authority. One domain, called speech, is explicitly denied as a subject for regulation. But what does this denial mean? It has long been established that the denial is not absolute. We have libel laws. We prohibit fraud in advertising. Somewhere lines are drawn in the grey area where speech shades over into action. Is picketing speech? Is parading speech? Is spending money on campaigns speech? These are the kinds of lines that Courts have to draw.

The first line that came to be drawn was a procedural one—the ban on prior restraint. A person can break the law by saying things; if he does so he can be sued in Court. But, so said the lawyers in the dawn of our country, what the First Amendment prohibited Congress from doing was setting up censorship, requiring a person first to submit what he proposed to say, so that its legality may be reviewed.

In the domain of speech, the state must suffer the disadvantage (so our forefathers concluded), of waiting till unrestrained speech was expressed, and then, but only then, if the outcome was illegal, to bring the speaker into Court.

That was a consensus among Constitutional lawyers. Indeed in the 19th century, conservative jurists who wished to limit the First Amendment argued that it was nothing but a ban on prior restraint. Congress, these conservatives argued, could prohibit various kinds of evil statements, but the one thing they conceded it could not do was to set up a review before publication.

Justice Parker in *Commonwealth v. Blanding* (3 Pick 304, 15 Am. Dec. 214) asserted the whole meaning of the First Amendment to be "to prevent all such previous restraints upon publication."

Today in the 20th century (since the days of Holmes and Brandeis, Douglas and Black), we have defined First Amendment rights much more broadly, but the principle of no prior restraint remains at the heart of the bundle.

True, the prohibition on prior restraint has never been absolute. Some, but not all lawyers have argued that there could be dangers so overwhelming as to justify the setting up of prior review and the licensing of speech in some domains. The First Amendment absolutists do not agree, but others argue that freedom of speech is just one consideration to be balanced with others. That is an issue among constitutional lawyers. So be it.

But what are the kinds of cases asserted to be so horrendous as to be exceptions? The most important case in which the courts have had to grapple with the problem, and in which in the end they came down on allowing publication, was the Wisconsin *Progressive* case, brought about when that magazine proposed to publish information on how to build an H-bomb.

That is an issue on which reasonable men could perhaps disagree. I am ready to listen to an argument that the dangers of that publication were sufficient to justify some quite exceptional measures. I do not know if that is true, but it is at least arguable. To worry about the diffusion of thermonuclear knowledge is not absurd.

But to put the harms done by a sociologist asking questions into the same exceptional category is absurd. There is no denying that we can all do harm by the questions we ask and the things we say. We can offend, we can embarrass, we can annoy. But these are precisely the harms that 200 years ago the framers of the Constitution decided had better be suffered rather than controlled. To require prior approval before a social researcher engages in an interview is as patently unconstitutional as it would be to require it of a reporter or of any ordinary citizen. To any student of constitutional history that is clear.

How remote those ideas of constitutional lawyers are from those of other intelligent people who do not share the guild acculturation of the legal profession first came to my attention some years ago when I served on a committee at MIT to review the procedures used by our IRB. It was my first exposure to the problems. When I learned about the practice of reviewing informal interviewing and library research by the IRB it was obvious that that was prior restraint on speech and quite unacceptable under United States law. The issue seemed to me obvious on the face of it. I wrote up a short memorandum and explained the issue to the committee, assuming that there would be no disagreement. How wrong I was!

Some of my colleagues—including behavioral scientists—were utterly puzzled by the whole argument, as was the chairman of the IRB. How could the IRB learn about the risky projects if they received no form describing the projects to be done. Granted, they would approve the non-risky projects quickly and without fuss, still, they asked: how could they do their job if no project descriptions were submitted to them.

The idea that some ancient lawyers and philosophers had concluded that in certain fields the burden of ascertainment had to be put on the enforcer seemed absurd to them. Making enforcement hard by saying one had to wait to discover a violation that had already occurred, rather than clearing up the matter by consulting in advance seemed to them to be doctrinaire foolishness.

We came out with a good majority report, but certainly no consensual understanding of the philosophy of the historical battle against censorship. There was no shared understanding of the notion that speech and action should be controlled by different standards; that the ethical norms of the medical profession might be different from the ethical norms of social investigation; that all harms are not the same; and that the harms imposed by verbal conflict are the price of a free society. The notion familiar to lawyers, economists, and social scientists that adversary combat is socially desirable and the injuries suffered in the process socially useful, was puzzling.

What was familiar to them were the canons of a different guild—the biomedical one. That is where the whole system of human subjects review began. It started with a 1963 incident in which Sloan-Kettering researchers injected live cancer cells into patients for an experiment. This led the United States Public Health Service to seek ways to protect itself from scandalous behavior by its grantees. When it made grants, the Public Health Service required the recipient institutions to set up review boards to check that subjects in the experiments it was supporting not be subjected to untoward risks.

Then in the late 1960s and early 1970s application of the procedures was extended form Public Health Service grants and contracts to grants and contracts by any part of the Department of Health, Education, and Welfare.

By that time, "harm" to human subjects had been construed to mean not only physical harm but "psychological" and "social" harm as well. Possible breaches of confidentiality were increasingly treated as justifying Institutional Review Board control, as much as did risks of bodily harm.

After 1974 the rules were broadened in still another way. HEW's obligation to protect human subjects was interpreted as requiring the Department to withhold grants from institutions that mistreated human subjects, whether or not the abuse took place in research the HEW was supporting. Thus, as in affirmative action cases, HEW-aided institutions were told that they had to apply IRB procedures to all research on human subjects, not just to research done under HEW grants.

By 1974 there was enough confusion as to just what the scope of the regulations was, or what it should be, that Congress mandated the establishment of a National Commission for the Protection of Human Subjects of Biomedical and Behavioral Research.

The Secretary appointed to the Commission three M.D.s, three lawyers, a physiological psychologist, a behavioral biologist in a medical school, a professor of bio-ethics at a medical school, and a professor of Christian ethics. No social scientists were included, a clue perhaps to what the Secretary assumed was the Commission's field of action. But the Commission saw its task in grander terms. Unrestrained by humility or wisdom, the Commission chose not to limit itself to protecting subjects from the medical risks with which the whole discussion began, but instead set out to assure the "ethical" conduct of research. Furthermore, it seemed obvious to the Commission that there should be no double standard. The ethical principles that applied to HEW-sponsored research should be applied to all research using human subjects. The Commission also concluded that harm was not just a medical matter; researchers could also do psychological and social harm to their subjects. Accordingly IRBs should not be limited to biomedical or experimental research but should also review social research. The Commission showed no awareness of the fact that the risks involved in social research are fundamentally different from those in biomedical research.

The most frequent risk to subjects of social research is that information about them which they would rather keep secret may become known to others. A risk of bodily harm is something quite different. Biomedical research often places the subjects at risk of physical harm, and possibly even death. Social science research causes harm to subjects by breach of confidentiality, embarrassment or exposure. To treat embarrassment by breach of confidentiality by the same set of regulations as are designed to control physical harm can lead only to absurdities and chaos. Breach of confidentiality can, of course, under some circumstances be an actionable matter. That occurs when there is an explicit (or sometimes implicit) contractual obligation of a person not to reveal what has been told him. Proprietary information is often given under pledge of secrecy. Pollers typically assure their respondents that their answers will be kept confidential; violation of such a promise could create a legal claim.

That is an ethical concern, and social science professions take it very seriously. They recognize an obligation to their informants to protect their confidences. But for the rest their relation to their subjects is not one of protection. It may indeed be the opposite.

In medicine or in clinical psychology the subject of research is often also a client. In these helping professions, though the potential harm done can be great, the researching professional may also be the subject's helper. The researcher may have a theraputic obligation to his client.

It is not so in journalism or in the social sciences. Their clients if they have any, are the public whom they inform, not the persons on whom they collect information. The HEW proposed regulations, however, tried to treat them all as though their ethical professional obligation was like that of a physician whose primary obligation is to do good to his patients. But the ethics of some of the social sciences, like journalism, may require practitioners to strip the concealing shroud from people whose behavior needs to be exposed for the good of others. Social scientists, like journalists, are engaged in what Sullivan was called "uninhibited, robust and wide open debate." Historians, political scientists, economists, and many sociologists are in that role. Their professional ethics are in no way inferior to that of the service professions; they are simply different.

Probably the majority of all social research examines situations in which people's behavior deviates from desired norms. As such it generally puts the persons studied at some risk of embarrassment and exposure. Whether the persons studied are brutal dictators, or corrupt politicians, insensitive employers or featherbedding workers, low achievement students or television viewers, whatever they may be they are people involved in some social problem, and so by definition not conforming to society's preferred ideals.

Social science could not exist if that kind of critique had to be defended at all times before committees whose main assignment was to take the side of the persons being studied.

It was to protect this practice of investigation, exposure, and criticism,—what the Supreme Court in the Sullivan case called robust, wide open debate—that censorship or prior restraint was prohibited in the domain of speech. Those who work in that style cannot live with a system of regulation designed for helping professions operating in areas of high risk.

THE NATURE OF THE UNIVERSITY

If constitutional doctrines about prior restraint were puzzling to those who did not come out of that tradition, the concept of the university has been even harder for modern bureaucrats to appreciate.

There is an anecdote, and I believe a true one, from the time when General Dwight D. Eisenhower became President of Columbia University. At this first meeting with the faculty the new president told them about various plans to do good things for the "employees" of the university. At the end the Nobel Prize-winning physicist, I.I. Rabi, rose and said: "Mr. President there is just one point: we are not the employees of the university, we are the university."

A university is not a bureaucratic structure of bosses and employees, but rather a loose confederation of independent scholars, each of whom decides for himself what he will study and how. The United States Supreme Court has reaffirmed that point in a different context; faculty members, the Court held, are not employees who must be bargained with collectively; they are autonomous decision-makers.

If faculty members are not employees subject to university control of their research, *a fortiori* students are not. They pay to be members of the university; they are not paid by it. They may be members of the university with the right to do research in it, but they are not controlled by it.

The concept of the university as a loose collectivity of independent scholars is an ancient one—a thousand years old. The university antedates the modern forms of bureaucratic organizations that emerged first in government in the 19th century and more recently in private corporations.

This new bureaucratic structure is so characteristic in our society that those who do not come from the university find it hard to understand that it is a very different kind of organization, and one that cannot direct its members.

Max Weber has given us the classic description of a bureaucracy, converting some Marxist concepts to the realm of information. The bureaucrat, like the proletarian, comes to work in an institution that he does not own, and where he owns none of the files, papers, or other means of operation. There he is part of a hierarchy of command, carrying out policies set at a higher level, and instructing those below him.

As Weber stressed, a bureaucracy can be a marvelously effective organization. It can have enormous power. But it is the power of the organization, not the power of the individual.

Compared to the notable in the public sector, or the entrepreneur in the private sector, whom the bureaucrat has replaced, this new man in the grey flannel suit, is a far weaker, less autonomous character. The power of the bureaucracy is in gearing men, whom individually it reduces, into a powerful collectivity.

A university is a more feudal and more entrepreneurial type of organization that preserves the ego and the autonomy of its individual members far more. As in the case of the First Amendment issues that we discussed above, the condition for that privilege which scholars enjoy is that, unlike bureaucrats in large organizations, they have very few resources of power besides their words. They do not control much by way of money or cadres. So they can be allowed to continue to enjoy far more autonomy than an employee.

That is the essence of a university: it is a collection of autonomous, independent scholars, deciding for themselves what they will do, and exercising an uncontrolled creativity. Our bureaucratic society needs that kind of organization, so out of keeping with its normal operation.

The human subjects regulations are not designed for that kind of organization. They are predicated on the assumption that the university as an institution has responsibility for what goes on in its premises. They presume that the university can instruct its students and faculty what research they may or may not do!

Those who favor such regulations have a point that they sometimes make. The university they say has changed. The university that I have been describing is already, they say, long since dead. Today they say, the university is a collection of bureaucratic teams, conducting funded research on contracts and grants, with principal investigators directing their staffs.

Yes, it is true that universities have absorbed much of this bureaucratic productive type activities. Medical schools are, above all, the place where this change has gone the furthest. They run hospitals, feed and care for patients, have hundreds of nonacademic employees, and have a classic bureaucratic hierarchy. There is much of it elsewhere in universities too.

I am not against such collective group research. It is what I do myself. For thirty years now I have always had in my title director or assistant director of some research program or other. Social scientists too engage in big science that takes teams to carry out.

But if we accept the notion that that is all that is left in universities, then the university is indeed dead. Its creativity, its intellectual contribution requires the careful nurturing within it of that sector that preserves a turf for the uncontrolled individual scholar.

Granted, sponsored research puts a responsibility on the institution and its central administration for accounting for the money received, and for fulfillment of the terms of the contract. But side by side with that activity, in any university worthy of the name, their resides still the ancient university which cannot even know what its members are doing with their time, much less controlling it. If the HEW-proposed regulations had ever been implemented, that would have been the end of the university as it has been for 1000 years.

I am not asserting that the human subjects regulations if they had been enacted would have been enforced very repressively. I do not believe so. They would have completed the process of bureaucratizing controls over virtually all university research. That is not the same thing as saying the bureaucratic controls would have been harsh.

Among social scientists, voices of moderation have suggested that a system of IRB censorship would not hurt because in the end they themselves would have run it. That is true. Just that has happened to a large extent in biomedical research. IRBs, when surveyed by the National Commission for the Protection of Human Subjects of Biomedical and Behavioral Research, were found more often to be deficient in favoring their medical constituency, rather than deficient in being too restrictive on them.

The same would surely happen in the social sciences. Senior establishment social scientists working the main stream would have had no trouble at all getting their activities approved. That would allow them to see the whole thing as no more than an administrative formality. The regulated would have captured the regulatory mechanism.

Yet "capture" is only a partial description of what happens in such circumstances. Graduate students and assistant professors, dissidents, and investigators of genuinely upsetting social issues would be the ones to find themselves censored. A simple capture theory neglects the fact that it is only a core elite group within an industry or profession that consumates the capture. The variety and diversity of tendencies that characterizes an unregulated profession gives way to discipline by a conformist center.

That is how a bureaucratic organization functions. It controls its members to assure effective implementation of the policies set at the top. That is not the way a university functions.

Let me close by asking what lessons we can draw.

The first lesson is a very practical one. I would urge on those of you who are in universities to make sure that when your institution submits its "general assurance" it does not promise to subject all research to the IRB process. There is research that carries risks and that should be so treated. The American Council on Education has drafted language that it is urging on the universities which defines the kinds of research which should be subjected to IRB review. Universities should use that language in their general assurances. It promises that review will be required for research that involves intrusion on the subject's person (the medical case) withholds usual or necessary resources from the subject (certainly a class of actions not protected by the First Amendment), or which uses deceit.

For the rest, if our universities are to remain universities, no one should ask to review and authorize what a scholar does in unsponsored research.

The second lesson I would draw is a theoretical one. As insitutions grow, and as centralized decision making increases, we have much to fear from failure to understand the principles which specialized cadres have created with their unique perspectives.

For those committed to freedom there is always much to fear from recurrent attacks. We have all heard that "eternal vigilence is the price of liberty." It is not a slogan that I like very much. If one has to be eternally vigilant one is not very free. To be free is to be able to relax and assume that no one will take one's rights away. Unfortunately it is not that way. I am afraid that it is true; we do have to be eternally vigilant. The freedom to relax and assume that our freedoms are safe is one we do not have.

Approval Voting: A Better Way to Elect a President?[a,b]

STEVEN J. BRAMS

Department of Politics
New York University
New York, New York 10003

INTRODUCTION

In this essay I shall discuss an election reform called *approval voting.* It appeared *de novo* about six years ago and has rapidly achieved considerable prominence. It now is the subject of over forty scholarly papers and has also been discussed by the media (see REFERENCES). I shall begin by commenting on the problem posed by multicandidate elections (those with more than two candidates), suggest how approval voting would ameliorate this and related problems of elections, show, as an illustration of its effects, how the results of New Hampshire's 1980 presidential primaries probably would have changed under approval voting, and finally indicate the reform's prospects for adoption. Although I shall devote a good deal of space to presidential elections, the problem of selecting a winner in multicandidate elections is a general one, which I shall illustrate with examples from nonpresidential elections.

THE PROBLEM

It is a sad fact that plurality elections (top vote-getter wins) are often rigged against majority candidates. For example, James Buckley was elected a United States Senator from New York in 1970 with only 39% of the vote. Buckley, a conservative, was probably no more representative of New York state voters than John Lindsay, a liberal, was representative of New York City voters when he was reelected to a second term as mayor of New York in 1969 with 42% of the vote. The 61% majority that voted for Buckley's two opponents, and the 58% majority that voted for Lindsay's two opponents, were effectively denied representation, at least in terms of their apparent ideological preferences, for six and four years, respectively. Moreover, as these examples indicate, the minority-candidate bias of plurality voting is not ideological: either liberals or conservatives may be aggrieved by the results.

One can think of many examples in which three or more candidates in a plurality contest divided the total vote such that no candidate received a majority. A dramatic recent example is the 1977 New York City mayoral

[a]This paper was delivered at a meeting of The Section of Science and Public Policy of the New York Academy of Sciences on January 7, 1981.

[b]This article is based in part on Brams[1,2]; it also contains data from the 1980 elections.

0077-8923/83/0403-0111 $01.75/2 © 1983, NYAS

election, in which six candidates got between 10 and 20 percent of the votes in the Democratic primary. This contest was followed by a runoff between the top two vote-getters, Edward Koch (19.8 percent in the plurality election) and Mario Cuomo (18.6 percent), which Koch won.

The runoff, however, offered no assurance that one of the four other candidates in the 10%–20% range could not have beaten Koch. Koch may well have been the strongest candidate in the Democratic field, but the plurality election, even followed by a runoff, did not offer incontrovertible evidence that he indeed was.

Neither did the three-way United States Senate election in New York in 1980 prove that Alphonse D'Amato was a stronger candidate than Elizabeth Holtzman and Jacob Javits, whom he defeated with a 45 percent plurality to 44 percent for Holtzman and 11 percent for Javits. In fact, an ABC News exit poll of voters (taken as they left the polling booths) revealed that if Javits had not been in the race, twice as many of his supporters would have opted for Holtzman as for D'Amato, easily making Holtzman the winner in a two-candidate contest. Ironically, the Liberal Party nominee (Javits) helped defeat the most liberal of the three Senate candidates (Holtzman).

The problem in all these cases is that a candidate, with support from a relatively small percentage of the electorate, can either win a plurality election outright or qualify for a runoff. When four-fifths of the electorate vote for other candidates, as happened in the 1977 New York mayoral race, the winner can hardly claim a great popular mandate. Worse, if there is another candidate whom a majority would actually prefer, the election system, in my opinion, is working perniciously.

This was probably the case in the elections that Buckley and Lindsay won in 1970 and 1969, respectively (as well as in the most recent Senate election in which Javits played the spoiler role). If either Buckley or Lindsay had been in a head-to-head contest with just one of his opponents, he almost surely would have lost to him.

What a shame for the voters when a winner in an election is in fact a weaker candidate than one or more of the losers! This paradoxical result most often occurs when two or more moderate candidates split the centrist vote, allowing a more extremist candidate to eke out a victory with only minority support. It is precisely this perversity of our system that seems in part responsible for the nomination of Barry Goldwater by the Republican Party in 1964, and the nomination of George McGovern by the Democratic Party in 1972. Although each nominee had vociferous minority support within his party, each was a disaster to his party in the general election.

A SOLUTION

I think there is a simple, practicable solution to this defect in our election system. It is a new voting system called approval voting, which allows a voter to vote for, or approve of, as many candidates as he or she wishes in a multicandidate race.

Thus, if there were five candidates, voters would not be restricted to voting for just one candidate. They could vote for two or more candidates if they had no clear favorite. However, only one vote could be cast for every approved candidate—that is, votes could not be cumulated and several cast for one candidate. The candidate with the most approval votes would win.

Approval voting has several compelling advantages over plurality voting, or plurality voting with a runoff:

1. It gives voters more flexible options. They can do everything they can under the present system—vote for a single favorite—but if they have no strong preference for one candidate, they can express this fact by voting for all candidates they find acceptable. For instance, if a voter most preferred a candidate who had little chance of winning, that voter could vote for both a first choice *and* a more viable candidate without worrying about wasting his or her vote on the less popular candidate.

If voters had had this option in 1980, a *Time* poll in mid-October, approximately three weeks before the presidential election, showed that 49% found John Anderson "acceptable"—and presumably would have voted for him under approval voting—versus acceptability figures of 57% for Jimmy Carter and 61% for Ronald Reagan. In other words, John Anderson would have been almost competitive with the major-party nominees under approval voting; had he been perceived as such, he surely would not have been as severely hurt by the wasted-vote phenomenon. (Anderson actually received 7% of the popular vote, to 41% and 51% for Carter and Reagan, respectively.)

Apart from the question of wasting one's vote, plurality voting is fundamentally unfair to a voter who may have a hard time deciding who is the best candidate in a crowded field but can choose the two or more candidates considered most acceptable. Approval voting thus provides the voter with more flexible options and thereby encourages a truer expression of preferences than does plurality voting.

2. It could increase voter turnout. By being better able to express their preferences, voters would more likely go to the polls in the first place. Voters who think they might be wasting their votes, or who cannot decide which of several candidates best reflects their views, would not have to despair about making a choice. By not being forced to make a single—perhaps arbitrary—choice, they would feel that the election system allows them to be more honest, which I believe would make voting more meaningful and encourage greater participation in elections.

3. It would help elect the strongest candidate. Today, as the previous examples illustrate, the candidate supported by the largest minority wins, or at least makes the runoff. Under approval voting, by contrast, it would be the candidate with the greatest overall support—the one most approved of—who would win.

An additional benefit is that approval voting would induce candidates to try to mirror the views of a majority of voters, not just cater to minorities whose votes could give them a slight edge in a crowded plurality contest, wherein a candidate with only 20% to 30% of the vote can win. The fact that

Jimmy Carter won the 1976 Democratic primary in New Hampshire with only 28% of the vote, and Henry Jackson won the primary in Massachusetts one week later with only 23% of the vote, says nothing about how acceptable either candidate was to the approximately three-quarters of the voters in each primary who did *not* vote for them.

4. It would give minority candidates their proper due. Minority candidates would not suffer under approval voting: their supporters would not be torn away simply because there was another candidate who, though less appealing to them, was generally considered a stronger contender. Because approval voting would allow these supporters to vote for both candidates, they would not be tempted to desert the one who is weak in the polls, as under plurality voting.

Such desertions occurred in the 1968 presidential election, when George Wallace dropped from 21% support in the polls in September to 14% in the actual vote on election day in November. It seems likely that the one-third of Wallace supporters who deserted him in favor of one of the major-party candidates did so because they thought he had no serious chance of winning—the wasted-vote phenomenon. But if there had been approval voting, Wallace would almost certainly have retained his original supporters, some of whom would also have voted for Nixon or Humphrey. One of these candidates would have won, probably with more than 50 percent of the vote (Nixon received 43.4 percent in the election; Humphrey, 42.7 percent), but Wallace would have registered much more substantial support than he did. Hence, minority candidates would probably receive their true level of support under approval voting, though majority candidates would generally win.

5. It is insensitive to the number of candidates running. Because ballots may be cast for as many candidates as the voter wishes under approval voting, if new candidates enter it is still possible to vote for one or more of them without being forced to withdraw support from old candidates. Plurality voting, on the other hand, is very sensitive to the number of candidates running. As the number of contenders increases, the plurality contest becomes more and more one of who can inch out whom in a crowded field rather than who is the strongest and most acceptable candidate.

Thus, when Jimmy Carter won less than 30% of the vote in the New Hampshire Democratic primary in 1976, the significance of his victory could be questioned. This gave the media an opportunity to pass judgment, which they gleefully did. Approval voting, by contrast, by better revealing the overall acceptability of the candidates independent of the field in which they run, more accurately mirrors each candidate's genuine support and renders more meaningful the voter's judgment. A side benefit would be that the voters, not the media, would weigh in more heavily in the selection process.

6. It is superior to preferential voting. An election reform that is used in a few places in the United States (for example, Cambridge, Massachusetts) shares many of the advantages of approval voting. It is called preferential voting, and it requires each voter to rank the candidates from best to worst. If no candidate receives a majority of first-place votes, the candidate with the fewest first-place votes is dropped and the second-place votes of his supporters are given to the remaining candidates. The elimination process continues, with lower-place votes of the voters whose preferred candidates are eliminated being transferred to the candidates that survive, until one candidate receives a majority of votes.

There are several practical and theoretical problems with preferential voting. First, it is difficult and costly to implement, and many voters do not understand how it works. For example, in New York City school board elections, in which this system has recently been used, it takes a week to count the ballots, make the transfers, and determine the winners. This delay could be alleviated by computerizing the system, but the capital expenditure needed to do this would be very substantial. Moreover, it seems to be a system that voters do not comprehend very well, at least as measured by extremely low turnouts of less than 10% in school board elections.

Second, under preferential voting, the candidate with the most first-place votes may be displaced after the transfers have been made to determine the majority winner. This may greatly upset that candidate's supporters, particularly if they are a large minority, and lead to questions about the legitimacy of the system—and ultimately its rejection by voters, as occurred in 1976 in Ann Arbor, Michigan. This challenge cannot be mounted against approval voting since approval votes are indistinguishable: whether these votes are first-place, second-place, or whatever is not recorded, so no portion of the winner's total can be judged "inferior."

Third, and more serious, preferential voting may eliminate the candiate acceptable to the most voters. For example, Goodell would have been eliminated as low man in the 1970 United States Senate race in New York, though there is evidence he could have beaten each of his opponents in face-to-face contests and would have won under approval voting.[3]

Fourth, and probably most serious, preferential voting violates a fundamental democratic ethic: a candidate may do better under this system by getting fewer, rather than more, votes. For example, a candidate can win an election when, say, 10,000 people vote for him or her but lose it when an extra 5,000 votes are added to his or her total. In other words, it may be in the interest of a candidate to tell some or his or her supporters not to vote for him or her to ensure a victory—a truly perverse result. Without going into details, this perversity arises not because additional votes are counted against a candidate but rather because of the successive eliminations of the ostensibly weakest candidates and the transfer of their votes to ostensibly stronger candidates.

Because of this perversity, preferential voting is said in the technical literature to violate a "monotonicity condition."[4–6] It is a pathology that does not afflict approval voting, because the more approval votes a candidate receives relative to the other candidates, the better he or she does. Under no circumstances can a candidate be hurt under approval voting by receiving additional votes, which seems to me a fatal flaw of preferential voting.

7. It is eminently practicable. Approval voting can readily be implemented on existing voting machines. With paper ballots, the counting of votes would be somewhat more tedious, but this should not, in my view, be a major barrier to its implementation.

To enact, approval voting will require a statutory, not a constitutional, change in most jurisdictions. Consider the statute in New Hampshire for voting in the presidential primaries. To enact approval voting in these elections would require only the substitution of the words in parentheses for the preceding words:

> Every qualified voter . . . shall have the opportunity . . . to vote his preference (*preferences*), on the ballot of his party, for his choice (*choices*) for one person (*any persons*) to be the candidate (*candidates*) of his political party for president of the United States and one person (*any persons*) to be the candidate (*candidates*) of his political party for vice president of the United States . . . .

It may be thought that, even given the virtues of approval voting, it would make little difference in a real election. This is so because the candidates would encourage voters to vote just for themselves (bullet voting) to keep down the vote totals of their opponents. Yet, even if the candidates made such an appeal, it would probably not be effective, particularly in a crowded race in which voters had difficulty distinguishing their single favorite. As evidence to support this contention, in an approval voting experiment involving several hundred Pennsylvania voters prior to their 1976 primaries, 72% of the voters voted for two or more of the eight candidates listed on their sample ballots.[7]

The commonsensical reasons I have outlined for adopting approval voting are supported by a number of technical arguments, which I shall just mention in passing for those who want to pursue the burgeoning mathematical and empirical literature on this reform. Among all nonranked voting systems, approval voting is more "sincere," "strategy-proof," and has the greatest propensity to elect so-called Condorcet candidates—those who, if they exist, would defeat all others in a series of pairwise contests[3,8–11]—and these results carry over to the election of committees;[12] it maximizes the "efficacy" of voters and is, in a particular sense, more equitable than other nonranked systems;[13,14] it has several advantages over preference, or ranking, systems, including a safeguard against the premature elimination of Condorcet candidates.[15–19] In addition, computer simulations, based on thermometer-scale ratings of real candidates by respondents in voter surveys,[20] experiments,[7] reconstructed ballot data,[16,19] and pilot survey data[21] lend empirical support to several of the postulated and derived theoretical effects. A number of theoretical contributions, particularly related to the strategic character and quantitative effects of approval voting and related systems have been made by Fishburn,[22–24] Hoffman,[25–27] Merrill,[28–30] and Weber.[31–33] Independently, in an extended analysis, Morin[34] has analyzed a number of features of approval voting, and Ottewell[35] has briefly discussed some numerical aspects of the system.

A CASE IN POINT: THE 1980 NEW HAMPSHIRE PRIMARIES UNDER APPROVAL VOTING

The results from New Hampshire's 1980 presidential primaries are now familiar. On the Republican side, Ronald Reagan was the big winner with

50% of the vote, with George Bush and Howard Baker trailing far behind with 23% and 13%, respectively. No other Republican candidate received more than 10% of the vote.

Would these results have changed significantly under approval voting? To approximate approval voting, ABC News asked voters in exit polls in New Hampshire to indicate *all* the candidates they considered acceptable if they could have cast more than one vote. The responses indicate that Reagan would have increased his total by 8 percentage points to 58% percent, Bush by 16 percentage points to 39%, but by far the biggest gainer would have been Baker: he would have climbed a dramatic 28 percentage points to finish in second place, behind Reagan, with 41% of the vote.

In other words, under approval voting Howard Baker would have more than tripled his plurality vote total and wound up a very creditable second-place candidate instead of finishing a dismal third. Although not the first choice of most Republicans, he had greater residual support than his closest rival, George Bush, on election day in New Hampshire. Unfortunately for Baker, the returns never revealed this fact, and he dropped out after second poor showings in the Massachusetts and Vermont primaries one week after the New Hampshire primary.

On the Democratic side, the order of finish probably would not have changed under approval voting. Even if Edward Kennedy had been approved by all Jerry Brown's supporters (10%), this still would not have been sufficient to wipe out Jimmy Carter's 11% lead (49% to 38%). To be sure, some of Carter's supporters would also have approved of Kennedy, and vice versa, but it seems very unlikely that Kennedy would have defeated Carter under approval voting. In fact, the ABC News approval voting figures confirm this proposition: Carter would have received 60% approval, Kennedy 48%, and Brown 36%.

In sum, the results reveal a more tightly packed race in the Republican primary, with one candidate displacing another, but no displacement would have occurred in the Democratic primary. To me the most important finding is that the race in both parties would have been more competitive and, in the Republican Party, also looked significantly different for one candidate (Baker), generally considered a "moderate," coming out of New Hampshire.[36]

PROSPECTS FOR ADOPTION

I believe that approval voting will be the election reform of the 20th century, just as the Australian, or secret, ballot—printed by the government with the names of all authorized candidates—was the election reform of the 19th century. In effect, the principle of "one person, one vote" under plurality voting becomes the principle of "one candidate, one vote" under approval voting. That is, each voter makes judgments about every candidate under approval voting, so the tie-in of a vote is not to the voter but rather to the candidates.

As an avowed advocate of general voting, I have campaigned for the adoption of this reform in the presidential primaries of New Hampshire and other states. (It was not used in any 1980 primaries, but it is now under discussion in several states, possibly for use in their 1984 presidential primaries; bills to enact approval voting have been introduced in the legislatures of New York and Vermont.) This campaign has received coverage in the *New York Times*[37,38]—wherein an Op-Ed essay also appeared[39]— the *Los Angeles Times,*[40] and several other newspapers, as well as on radio and television. Editorials have also appeared,[41] and the reform has been discussed in such magazines as *Scientific American.*[42]

Over the past few years, I have had voluminous correspondence from presidential candidates, Secretaries of State, state legislators, state chairpersons of the Democratic and Republican parties, major public interest organizations such as the League of Women Voters and Common Cause, and others about the adoption of approval voting. Approval voting is receiving serious consideration in about a dozen states now.[c] At the local level, a University of Pennsylvania group recommended, in a commissioned study, that approval voting be used in nonpartisan elections in Atlantic City, New Jersey, in a proposed charter revision for that city.[44] A sourcebook and synthesis of work on approval voting through 1982 has recently appeared.[45]

Many voters are cynical today. Their choices are artificially limited. Approval voting could go a long way to defusing their anger and frustration at a system that has nominated patently unrepresentative candidates like Barry Goldwater and George McGovern, who destroyed their parties' presidential hopes in the years in which they ran and gave many voters no viable alternative to their opponents. By averting the nomination of such candidates, approval voting would strengthen the parties by tending to elect their strongest nominees and increasing competition in the general election. Moreover, by letting voters express themselves better at the polls, approval voting would greatly enhance democratic choice. This essentially costless reform thus seems well suited for voters, parties, and the heightening of competition among candidates that maximizes a voter's viable options.

REFERENCES

1. Brams, S. J. 1979. Approval voting: A practical reform for multicandidate elections. National Civic Review **68:** 549–553, 560.
2. Brams, S. J. 1980. Approval voting in multicandidate elections, Policy Studies Journal **9:** 102–108.
3. Brams, S. J. & P. C. Fishburn. 1978. Approval voting. American Political Science Review **72:** 831–847.
4. Smith, H. J. 1973. Aggregation of preferences with variable electorate. Econometrica **51:** 1027–1041.

[c]Expressions of support for approval voting from a number of individuals are quoted in Brams.[43]

5. DORON, G. & R. KRONICK. 1977. Single transferable vote: An example of a perverse social choice function. American Journal of Political Science **21:** 303–311.
6. FISHBURN, P. C. 1980. Monotonicity Paradoxes in Elections. Discrete Applied Mathematics **4:** 119–134.
7. KELLETT, J. & K. MOTT. 1977. Presidential primaries: Measuring popular choice. Polity **9:** 528–537.
8. BRAMS, S. J. 1976. One Man, n Votes. Module in Applied Mathematics, Mathematical Association of America. Ithaca, NY.
9. BRAMS, S. J. 1977. When is it advantageous to cast a negative vote? *In* Lecture Notes in Economics and Mathematical Systems (Mathematical Economics and Game Theory: Essays in Honor of Oskar Morganstern) **141.** R. Henn & O. Moeschlin, Eds. Springer-Verlag. Berlin.
10. BRAMS, S. J. 1978. Comparison Voting. Innovative Instructional Unit, American Political Science Association. Washington, DC. Published in slightly revised form in Modules in Applied Mathematics: Political and Related Models, vol. 2, S. J. Brams, W. F. Lucas & P. D. Straffin, Jr., Eds. Springer-Verlag, New York, 1983.
11. BRAMS, S. J. 1978. The Presidential Election Game. Yale University Press. New Haven, CT.
12. FISHBURN, P. C. 1981. An Analysis of Simple Voting Sytsems for Electing Committees. SIAM Journal on Applied Mathematics. **41:** 499–502.
13. FISHBURN, P. C. & S. J. BRAMS. 1981. Efficacy, power and equity under approval voting. Public Choice. **37:** 425–434.
14. FISHBURN, P. C. & S. J. BRAMS. 1981. Expected utility and approval voting. Behavioral Science. **26:** 136–142.
15. FISHBURN, P. C. & W. V. GEHRLEIN. 1982. Majority Efficiencies for Simple Voting Procedures: Summary and Interpretation. Theory and Decision **14:** 141–153.
16. BRAMS, S. J. & P. C. FISHBURN. 1981. Reconstructing voting processes: The 1976 House majority leader election under present and alternative rules. Political Methodology. **7:** 95–108.
17. FISHBURN, P. C. 1980. Deducing majority candidates from election data. Social Science Research. **9:** 216–224.
18. FISHBURN, P. C. & S. J. BRAMS. 1981. Approval voting, Condorcet's principle, and runoff elections. Public Choice. **36:** 89–114.
19. FISHBURN, P. C. & S. J. BRAMS. 1981. Deducing simple majorities from approval voting data. Social Science Research. **10:** 256–266.
20. KIEWIET, D. R. 1979. Approval Voting: The Case of the 1968 Election. Polity. **12:** 170–181.
21. BRAMS, S. J. & G. SHARRARD. 1979. Analysis of Pilot Study Questions on Preference Rankings and Approval Voting. Mimeographed.
22. FISHBURN, P. C. 1978. Axioms for Approval Voting: Direct Proof. Journal of Economic Theory. **19:** 180–185.
23. FISHBURN, P. C. 1978. A Strategic Analysis of Nonranked Voting Systems. SIAM Journal of Applied Mathematics. **13:** 488–495.
24. FISHBURN, P. C. 1978. Symmetric and Consistent Aggregation with Dichotomous Voting. Aggregation and Revelation of Preferences. J. J. Laffont, Ed. North-Holland. Amsterdam.
25. HOFFMAN, D. T. 1982. A Model for Strategic Voting. SIAM Journal of Applied Mathematics. **42:** 751–761.

26. HOFFMAN, D. T. 1979. Mathematics and Voting: Approval Voting Is Better. Mimeographed.
27. HOFFMAN, D. T. 1979. Relative Efficiency of Voting Systems: The Cost of Sincere Voting. Mimeographed.
28. MERRILL, S. III. 1979. Approval Voting: A "Best Buy" Method for Multicandidate Elections? Mathematics Magazine. **52:** 98–102.
29. MERRILL, S. III. 1980. Decision Analysis for Multicandidate Systems. UMAP Module, Unit 384. Education Development Center. Newton, MA.
30. MERRILL, S. III. 1981. Strategic Decisions under One-Stage Multi-Candidate Voting Systems. Public Choice. **36:** 115–134.
31. WEBER, R. J. 1977. Comparison of Voting Systems. Mimeographed.
32. WEBER, R. J. 1977. Multiply-Weighted Voting Systems. Mimeographed.
33. WEBER, R. J. 1977. Reproducing Voting Systems. Mimeographed.
34. MORIN, R. A. 1980. Structural Reform: Ballots. Vol. 1. Vantage Press. New York.
35. OTTEWELL, G. 1977. The Arithmetic of Voting. In Defense of Variety. **4:** 6–8.
36. BRAMS, S. J. 1980. Baker Could Have Survived N.H. *Concord* (N.H.) *Monitor.* March 8.
37. WEAVER, W. JR. 1979. New System Urged in Presidential Primary Voting. *New York Times.* April 13.
38. BROWNE, M. W. 1980. Can Voting Become Safer for Democracy? *New York Times.* June 1.
39. MERRILL, S. III. 1979. For Approval Voting. Op-Ed Page. *New York Times.* July 20.
40. LEMBKE, B. 1979. A Proposal for Legal Ballot-Box "Stuffing." *Los Angeles Times* (Orange County section, part 2). May 12.
41. BRAMS, S. J. 1980. Approval Voting in N.H. Guest Editorial. *Concord* (N.H.) *Monitor.* January 9.
42. GARDNER, M. 1980. Mathematical Games. Scientific American. October.
43. BRAMS, S. J. 1981. Approval Voting: One Candidate, One Vote. Archway. **2:** 10–14.
44. GOVERNMENT STUDY GROUP. 1979. A New Government for Atlantic City: A Strong Mayor—Strong Council Plan. Department of Political Science, University of Pennsylvania. Philadelphia.
45. BRAMS, S. J. & P. C. FISHBURN. 1983. Approval Voting. Birkhäuser. Boston, MA.